Desire in the Age of Robots and AI

"Rebecca Gibson has given us this wonderful book that details how these technologies could have an even more profound influence—redefining that in us which is most human."

—Chris Irwin Davis, *Ph.D., A.I. Research Scientist*

"Half of the time I plug into the world through technology I am wondering about the 'people' of the other side.... As the online space expands, we become increasingly less sure where we stand in relation to digital simulacrums. Gibson's book is a delightful and dirty look on what separates us from the machines, and exactly what those differences say about us if they even exist anymore."

—Jef Rouner, *author of The Rook Circle*

"Well researched, engaging, and thought provoking, Dr. Gibson has given humanity a lot to consider when it comes to our desires and creations. The dance between science fiction and advances in technology is always complex and uniquely intertwined, but the threads here are neatly pulled apart and examined in such an insightful way that the reader can't help but find new understanding of their own inner world and the desires inherent to the human condition. A winning combination of entertaining and well documented information and analysis!"

—Eli Girvin, *Talent Analytics Strategy Consultant with IBM*

Rebecca Gibson

Desire in the Age of Robots and AI

An Investigation in Science Fiction and Fact

palgrave
macmillan

Rebecca Gibson
Department of Anthropology
University of Notre Dame
Notre Dame, IN, USA

ISBN 978-3-030-24016-5 ISBN 978-3-030-24017-2 (eBook)
https://doi.org/10.1007/978-3-030-24017-2

This Palgrave Pivot imprint is published by the registered company Springer Nature Switzerland AG
The registered company address is: Gewerbestrasse 11, 6330 Cham, Switzerland

This book is dedicated to the cyborg in all of us,
and the ideas of a nautilus and a nice cup of tea.

Acknowledgments

The hard work of writing this book would not have been possible without my massive support network, which includes way too many people to mention here. Most notably, I must thank my parents, John and Judy Horwitz (you can stop sending me robot sex articles now, Dad, the book is published), for being the best parents ever; Linda Langley, who supported me during the first phase of this research and has continued to be my cheerleader ever since, for keeping body and mind together during that time; Jef Rouner, who is a true wordsmith, for inspiring me to reach for my better self; Jay VanderVeen, who has generously and tirelessly listened to me expound on various iterations of this project, for making sure I did not lose my mind into cyberspace; and Agustín Fuentes, who has given me wholehearted personal and departmental support during the writing phase, for all the early-morning meetings and late-night rambling emails. Finally, I would like to thank my excellent editorial team, Mary Al-Sayed and Madison Allums, for their kindness, promptness, advice, support, and calmness in the face of my inexperience.

Contents

List of Figures

CHAPTER 1

The Gears and Wires of Robot Sex

Abstract This chapter introduces the main theme—that sex robots have multiple uses; they fulfill our desire to create, they are physical manifestations of our storytelling natures, and they are there to be the perfect woman (and they usually are, indeed, female in form). In this chapter, Gibson unpacks some of the reasons we feel the need, as humans, to explore robotics, cybernetics, and cyborgs, rather than sticking to non-physical/fictional, but fully human, companions, and then looks at why we desire our creations and why that desire morphs into fetishization—which often has a racialized component. Finally, Gibson discusses the idea that sci-fi is both predictive and reactive, and we tell stories to explain our desires to ourselves and others.

Keywords Robot sex • Storyteller • Science fiction • *Do Androids Dream of Electric Sheep?* • *Blade Runner* • *Blade Runner 2049*

Introduction

Robot sex. The act of sexual intercourse by a human with a robotic or artificial intelligence type being. Yet, somehow, more than that as well. Intimacy, romance, courting, coitus, oral sex, companionship, and the fundamental way we see ourselves have become interlinked with these two words. This book will examine how science fiction's portrayal of humanity's

R. Gibson, *Desire in the Age of Robots and AI*,
https://doi.org/10.1007/978-3-030-24017-2_1

desire for robotic and artificially created humanoid companions influences and reflects changes in our actual desires. I will begin by taking the reader on a journey through the theories that outline basic human desires—we are storytellers, and we need the objects of our desire to mirror many, though not all, aspects of our beings in order that the stories we tell ourselves match our realities. I will continue by looking at three variations on one story—the 'Blade Runner' mythos—starting with the original Phillip K. Dick novella *Do Androids Dream of Electric Sheep?* (1968/2017), and then moving to the two movies in the series "Blade Runner" (Scott, 1982) and "Blade Runner 2049" (Villeneuve, 2017). Using these fictional portrayals, this book will examine the stories we tell ourselves regarding our intimate relations with robots.

There are reasons we seek out differences in our mates; having something mirror us too closely when we know they are not the same makes our skin crawl. What, then, makes a created being—whether partial cyborg or completely mechanistic—human enough, but not too human? Science fiction both creates new species of potential companions and highlights what we already want and how our desires dictate and are in return recreated by what is written. Sex with robots is more than a sci-fi pop-culture phenomenon; it's a driving force in the latest technological advances in the cybernetic and artificial intelligence sciences. As such, this book fulfills a need to examine both what is in our minds and what is in our capabilities, when it comes to the creation of such beings. Contemporary scholarship does not seat this idea so firmly in the basis of science fiction, nor does it always ask the ethical questions about how we need to address the rights of and our responsibilities toward the new life forms we may soon engender. In the following chapters, I will attempt to move closer to a deeper understanding of those questions by examining the science behind the fiction and what that fiction tells us.

To give a short synopsis of the chapters, this chapter, that is, Chap. 1, will introduce the main theme—that sex robots have multiple uses; they fulfill our desire to create, they are physical manifestations of our storytelling natures, and they are there to be the perfect companion (though they usually are female in form). I will unpack some of the reasons we feel the need, as humans, to explore robotics, cybernetics, and cyborgs, rather than sticking to non-physical/fictional, but fully human, companions, and then look at why we desire our creations, and why that desire morphs into fetishization—which often has a racialized component. Finally, I will discuss the idea that sci-fi is both predictive and reactive, and we tell stories to explain our desires to ourselves and others.

In Chap. 2, I will show how the physical nature of the androids is such that they are excellent examples for this conceit—human skin over an organic-mechanic built frame, they are the ultimate cyborgs, with programmed mental/emotional natures and created bodies, who can, or are trying to, pass as human. In *Do Androids Dream of Electric Sheep?* (Dick, 1968/2017), the first iteration of the Blade Runner/Electric Sheep mythos, the division between human and non-human, is starkly clear—humans are born, made of flesh and blood, and have created the androids to do work they do not wish to do. However, their humanity is ebbing away, signified by the emotion control box, which is used to dial emotions up or down, and to implant specific desired emotions in their minds. The question then becomes as follows: Are the androids showing more humanity than the humans? This question leads us to another one: How human do robots have to be in order for us to find them desirable? In the book, these questions center on the relationship between a Replicant, Rachael, and the man who seeks to destroy her, Rick Deckard. Some answers come from recent developments in android design and the reactions to it.

Chapter 3 will examine the first *Blade Runner* movie, Ridley Scott's adaptation of the original novella (1982). In *Blade Runner*, Replicants become undetectable, undifferentiated visually, from humans. The newer models, personified by Rachael (modified somewhat from her original textual incarnation), can pass the Voight-Kampff test—a test which distinguishes true emotional responses from implanted ones. This perfect mimicry is our next iteration of how we desire sex robots to behave, and it is desired well in advance of the possibility of building such bots.

It is in this iteration of the story where two theoretical questions appear—that of the racialization of the characters, which is far more apparent in visual media, and the humanity of death—how being able to die, and the fear of it, humanizes the movie's Replicants. These will be addressed via various studies done both in hypothetical and concrete approaches to race and gender (Allan, 2015; Chu, 2015; Prater & Fung, 2015; Roh, Huang, & Niu, 2015). I will also address what would constitute 'death' for a modern-day sex-bot, or, conversely, what makes one alive (Bordo, 1987; Levy, 2007/2008; Smith, 2013). Is it mechanical? Biological? Physiological? How close are we getting to the human experience of birth, life, and death, and how does that closeness tie into how human-feeling the newest developments in AI and robotics are?

In the final descriptive chapter, Chap. 4, I will examine the most current iteration of the story, *Blade Runner 2049*, and tease out implications for the future of robotics. While the gender dichotomy is very much upheld in this movie, it is considerably more nuanced than in the first *Blade Runner*.

Throughout each iteration of the story, a thread of interconnectedness has stood out—the androids/Replicants are created by humans and need humans to have any sort of existence at all, but humans need the Replicants, both to do dirty work and as a source of identity. Humans identify themselves against a contrast of Replicant life in a very binary fashion. The creator is meaningless without the creation, for to know what it is to be human, there must be an inhuman correlate. For the latter part of this chapter, I will include interviews with robotics developers about their own desires, and why they chose to make hominoid robots. This, then, will be paired with an analysis of Rachael's reproduction—the fact that she and Deckard had a child organically in Villeneuve's movie (2017). What happens when the creator is replaced, the creation can give birth to life, and humanity can no longer claim that as a reason for human exceptionalism? Will our own robotics ever be able to self-replicate?

Chapter 5 will conclude this book and look at the way the stories are changing to be somewhat more inclusive, though certainly there remains a long way to go. It will begin by reiterating some of the themes from *Blade Runner 2049* (Villeneuve, 2017), including that of the potential for self-replication. Then it will examine the revolt of the sex robots in Ex Machina, while discussing the perils of giving something an emotive capability while forgetting that rage, hatred, and the will to survive are also emotions. It will also look at the diversity of the webcomic *Questionable Content* (Jacques, 2003–2019), and examine how it fits into the greater narrative.

Woven into this, the final chapter will also contain a look at the very newest of advancements in cybernetics and AI, examining biofeedback prosthetics, 'learning' AI, and following the latest news on Sophia, the Saudi citizen robot. I will continue to examine the testimonials of people who develop relationships with the sex dolls, including pushback from those who are skeptical about the utility of such relationships, and end the chapter with a discussion with the creator of the magazine *Strange Horizons* about how she is encouraging a more inclusive sci-fi.

Glossary

Before getting started, certain terms must be defined:

Robot A mechanical structure that performs certain tasks based on inputs from software and its own functional capabilities. Robots can be as simple as a long arm that is programmed to pick up a piece of material and transfer it to another location, or as complex as a human-shaped machine which can speak and respond to conversation in ways that mimic our own emotive states. In *Do Androids Dream of Electric Sheep?* (Dick, 1968/2017), they are referred to as androids or andys. In "Blade Runner" (Scott, 1982) and "Blade Runner 2049" (Villeneuve, 2017), they are referred to as Replicants.

AI Artificial Intelligence, or AI, is a software or programming that demonstrates human traits such as speaking, emoting, learning, intuiting, and reasoned problem solving. It can be used for such operations as automated telephone systems or to enhance a person's experience with a robot.

Cyborg A being with both cybernetic and organic parts—in the case of a human, prostheses which replace amputated limbs or repair damaged or missing bodily systems (such as cochlear implants), technically make that human a cyborg.

Cybernetic Having to do with the interface between animal and machine—where control and command meet.

Sex While the definition of sex or sexual intercourse is one that shifts based on the time and culture, for the purpose of this book it will be considered stimulation or penetration of the genitals by hands, mouth, other body parts (including other genitals), or a mechanical substitute.

Love Affection, caring, desire, and/or infatuation, as expressed from one being to another.

Science Fiction Versus Science Fact: The Stories We Tell Ourselves

The reality of science itself began hundreds of years ago, with Persian, Greek, and Roman experiments in the realms of alchemy and physics, and the Mesoamerican practice of electroplating. What all good science has in common are principles of hypothesis, experimentation, and replicability. A scientist has a question, uses experimentation to refine that question, and eventually finds an answer, whether in part or in whole, that can be repeated by disinterested parties. The field of science fiction, on the other hand, has existed for a considerably shorter period of time, with Mary Wollstonecraft Shelley's *Frankenstein: or, the Modern Prometheus* (1818/1831) being widely acknowledged as the first true sci-fi novel.

Robots and the idea of sex with robots, like both the story of Frankenstein and the eponymous monster, are an amalgam of science and sci-fi, a chimera of reality and the stories we tell ourselves about what reality could be. The original robots—automatons which operated by means of steam, levers, pulleys, or hydraulics—are as old as other scientific devices (Voskuhl, 2013), and have always embodied a feeling of the mystical, the fictional, the not-quite-real, as they mimic humans but are, fundamentally, not human at all. However, as the concept of the robot has progressed through time and melded with the technical advances of the twentieth and twenty-first centuries, the physicality of robots and the way they are discussed in popular culture and in the media are increasingly human-like, as is the skepticism about our own desires versus how we would realistically interact with them. To this end, one finds such tweets:

> Who the hell are y'all trying to fool with this 'sexbots will replace human women' shit anyway. I've been inside your houses, you can't even use a coaster properly. Now you're trying to tell me you'll clean and maintain a robot that you cum in. Okay josh. [Tweet by Jerk du Soleil @dubsteppenwolf [all spelling, punctuation, and capitalization sic.] 1:54 am, January 19, 2018].

As this book will explore, attitudes toward the ideas and the technology are varied.

One can see this as a recursive system, a circle that has no beginning and no end. Robotics inspires sci-fi to give us new permutations of stories, and those stories inspire scientists to turn the possible into the probable, and then into reality. It was only a matter of time before sex entered the equation. And not just sex, love.

In fact, love lies at the heart of one of the very first statue-made-life tales, that of Pygmalion and Galatea. In this ancient Greek myth, a sculptor creates a statue of a woman so beautiful that he falls in love with it. Showering the statue with affection, gifts, and prayers to Aphrodite, he has no desire to have anything to do with human women. His prayers were eventually answered, with Aphrodite breathing life into the inert marble, creating a beautiful, perfect, animated being. In this simple tale, we have several ideas which will be the central themes of this book—the desire (and power) to create, the desire for perfection of the physical form, the concept of a love that is eternally and perfectly returned to the giver, and the compelling nature of combining all those ideas into one created mechanical being.

Creation can be seen to be tied to procreation—to the reproduction of ourselves, and therefore our species. Yet many robots or androids shown in modern sci-fi are, even if referred to as the children of their creators, not shown as children or child-like at all. Therefore, a robot with human form usually enters into a primary duality—that of offspring and potential mate (if not for the creator, then for another). Dualities come into play consistently during discussions of robotics and of humans—it seems we find comfort in juxtapositions. It makes logical sense that creators would want humanoid robots, that is, robots created in their own image, to fulfill an entire human life cycle: birth, to life, to reproduction, to death. Yet, the physicality of robots precludes two of these: reproduction and death.

We can see the first two properties easily. A robot is born when its creator physically forms it into being and 'turns it on,' whether that means beginning the water pressure for its hydraulics or charging its batteries. It lives for the duration of time that it is active. Robots are inherently Cartesian (Bordo, 1987)—their minds and bodies are separate, and although there is an interface between the two, that interface is physical/mechanical, with no "mind" that is anything other than software-driven. And so, we define and control both the birth and the life. However, the Cartesian nature of robots makes the final two properties unrealistic.

Robots cannot reproduce outside of their internal programming—there are some robots that can replicate (Cornell, 2005; PopSci, 2014), but this is very much a product of their programming and of outside instructions. The robot cannot make the decision to reproduce of its own desires, nor, indeed, can it avoid reproduction even if it did not 'want' to create another robot. Human reproduction, despite being seen as a biological imperative by some, can be timed, planned, altered, or halted by the use of biochemical, surgical, or abstemious interventions. Humans have multiplicitous cascades of nuanced desires, fears, impulses, and drives, which they can indulge or refrain from indulging at will. Robots have programming. When a robot reproduces, it is obeying an external command—a command that another being, a human, wrote into its software. It is, at the moment, impossible for a robot to have (free) will.

So what about death? Surely if a robot can live, it can also die? Not necessarily. Humans over time have tended to perceive death as a binary state (Doughty, 2017), where you either are alive with all the attendant physical facets of aliveness, or you are dead and your physical being decays. Certain societies recognize a transitionary liminal phase where the body and/or soul is in the process of becoming dead, and some of those phases are longer than others, but it can still be reduced to the binary.[1] Once dead, most people understand that there is no (actual, physical, non-metaphorical) returning to life. The religious ideas of life after death are either metaphorical, or do not tend to consistently and predictably intersect with this plane of being. Death is a permanent state.

This permanent state does not apply to beings which are mechanically dualistic, that is, beings with replaceable, fixable, mechanical parts, which can be swapped out when broken, or where the software can be updated if necessary. An argument can be made that cybernetic advancements, such as biofeedback prosthetics, cochlear or retinal implants, pacemakers, or other biomechanical 'replacement parts,' are doing the same for humans. However, the difference is in the permanence of human death and the essential uniqueness of the individual human mind. Once brain and heart stop, and the body starts to decay, there is no recovering from (human) death, no transferring the brain to another shell, no reawakening of that singular personality, or memories, or mind into a new and working body. For a robot, software can be copied and transferred to new media, mechanics can be swapped for newer models, and that spark of 'life' can awaken again, in a new configuration with the old memories and ideas intact.

Stories Against the Dark

What is the use of dreaming of robots as indistinguishable from humans if the reality does not match the fiction? Claude Levi-Strauss (1962/1966) writes that humans are storytellers, and that the way we are distinguishable from our closest primate relatives and the way we perpetuate our culture are through fiction. We do not merely pass things down as they are, literal and spare; rather, we analogize, mythologize, expand, use rhetorical constructions, and create interlocking worlds of stories, to make our tales all the more memorable. He called this concept '*bricolage*' (ibid.) after the

French practice of improvisational Do-It-Yourself handiwork and house-building. The builder or handyman—the *bricoleur*—would take whatever was available from the immediate surroundings, paying little mind to planned fit, purposeful stylization, architectural symmetry, or deliberate aesthetics, and the resulting hodge-podge of construction would become something organically beautiful, something memorable.

This memorable quality, this beautiful patchwork, characterizes storytelling traditions, both oral and written. Levi-Strauss was making generalizations about observed trends in cultures around the world, and his observations hold up to scrutiny even today, as do those of mythologists Frazer, Colum, and Campbell (1950/1996; 1921/1990; and 1949/1973, respectively). All three mythologists, and Levi-Strauss, posit the idea that stories generally follow one of several available formats, tropes, or templates if you will, and the uniqueness and variation come from the details of the narrative, not the narrative form.

For example, the Hero's Quest trope begins with a hero who does not yet know he is a hero. He is given a challenge or becomes part of a conflict, for whatever specific reason, which induces him to leave home. He may or may not start out with a purpose, but once gone from home, he meets (someone), who complicates matters. At some point, his life will be in peril. This complication and peril eventually resolve through his choices, turning him into the hero, and he returns home triumphant, having bested (something). As you can see, considerable ambiguity exists in the unformed details of the story. I have used male pronouns, but the hero can be female or non-binary. I have said they meet (someone) and best (something) and that person and that thing change depending on what the person telling the story determines to be important. Does the hero need to slay a dragon? Does the hero need to wrestle with their inner demons? Does the hero meet a guide, or does he meet a foe? All are dependent on the particular motive of the *bricoleur* and the details they choose to include in the narrative.

There are five generally historically recognized conflict patterns in literature: (hu)man against nature, (hu)man against god, (hu)man against society, (hu)man against other (hu)man, and (hu)man against self. Each of these conflicts can be found in science fiction, in various permutations. Robots, however, do not seem to fit into these, and in fact they create a sixth, and considerably more modern pattern: (hu)man against technology. With technology advancing at an exponential rate, we are both its creators and its combatants. Many science fiction stories grapple with technology that gains a sentient understanding of its own circumstances and turns against its creators. That question of sentience truly creates the conflict. In all the original historical conflict patterns, the antagonist represents a type of living, thinking, breathing thing—even nature often ends up being portrayed as a sentient antagonist, with storms, earthquakes, deserts, and oceans given anthropomorphic form or abilities. Yet that sentience is innate, not created. It occurs spontaneously, and not as a result of programming. For humans to fight against, or love, technology, we need to create a being that 'feels' human in a way that makes us believe that we are not alone.

Are We Alone Together?

As mentioned earlier, the desire for an automaton, a robot, or created lover is nothing new. However, the ability to create one that reacts, responds, converses, and makes us feel less than alone is relatively new on the scale of available technology. So, what does that mean? What does that look like? Both Sherry Turkle and David Levy (2011 and 2007, respectively) agree that the primary component in a feeling of togetherness, or the absence of loneliness, is companionship, or a feeling that the being we are talking to, sitting with, making love with understands how we feel and empathizes with us. They also stress that the physicality of the being must come relatively close to human standards, with Levy (2007) citing Masahiro Mori's theory of the uncanny valley (pp. 159–160).

The theory of the uncanny valley maintains that humans are desirous of 'human-likeness' in robots increasingly up to approximately 80% of similarity. After that, the desire strongly drops off, as the robot becomes too 'uncanny,' too much like 'human with something wrong' to be considered desirable as a companion. We seem to want to fundamentally know that the being sitting in front of us is not human. One way we can 'know' this is the administration of the Turing Test. A Turing Test consists of a test of any number of parameters designed to determine if the person being tested is human or AI. The parameters can be physical, verbal, written, or biochemical (e.g., one markedly cruel test could be to administer pain, and see if cortisol in the 'blood' increases), but the import remains that a robot who can mimic humans enough that they fool the person administering the test has 'passed,' whereas a robot who cannot will inevitably fail. We seem to want our robots to fail the Turing Test, even if just a little bit.

In this way, we may be stymying our own desire for togetherness—if we wish our companions to be recognizably not-human, then we will never have true companionship. However, Levy identifies ten principle factors that impact whether or not we fall in love with our companion: similarity, desirable characteristics of the other, reciprocal liking, social influences, filling needs, arousal/unusualness, specific cues, readiness for entering a relationship, isolation from others, and mystery (2007, pp. 143–150). The right combination of these traits can overcome our feeling of uncanniness. These traits are all possible, to varying degrees, in current robotic technology. As they are significantly impactful to what we see represented in science fiction, and also what drives that science fiction, I will explore them one at a time.

Whether in facial muscle mimicry (New Scientist, 2008) or in biofeedback prosthetics, "similarity" (p. 144) (when not uncanny) gives us comfort—we can see the empathy in a facial expression similar to our own, or know that our limb has been replaced by something that does not feel or look too unusual to us. However, similarity does not just need to occur in appearance. We need similar weight in a human analog to convince us that we are together with something human like ourselves, similar speech patterns and, some argue, similar scents (Levy, 2007). This similarity always tempers itself against the idea of too much, though, and we become uncomfortable with, for example, automated phone systems which say things like "I'll look that up!" or "sorry, I didn't catch that" (Comcast, 2018) in a very cheery human-sounding voice.

This, then, leads into "desirable characteristics of the other" (Levy, 2007, p. 145). Levy lists these as falling into two categories, "personality and appearance" (ibid.), which make sense in terms of human sexual desire. We want someone who performs an agreeable personality while also tripping our trigger in terms of looks. In a companion we can essentially craft for ourselves, there exist very few reasons outside of cost to not create one which appeals to both. Here, the software and hardware can work together to create a total package—a sex-bot who looks like our dreams, and who we enjoy talking to and being with.

Even without an AI- or software-based personality, the available sex dolls of the first part of the twenty-first century often got personalities ascribed to them by their owners. For example, one anonymous testimonial from the RealDoll website, posted on November 23, 2016, states:

> Beautiful Muse Sabrina!! I recently received my gorgeous hybrid-doll (Sai) or as I renamed her Sabrina. I'm extremely impressed with her! The artistry that has gone into her creation, is beautiful. I'd have to say that Abyss Creations, and Boytoy Dolls have a very talented crew of artists in their employ! She is obviously just a doll, but for me she is much more. She is a presence, and even though she is inanimate, I love projecting and reflecting how I think and feel about her, when we are together. Intimacy is also phenomenal too. Loving her, in every way, is not at all disappointing! As an artist myself, I also plan on using her as a model for drawing and painting. I'm extremely pleased!! She is a 'RealDoll' for sure!! [RealDoll, n.d.].

The buyer is attracted not only to her physical being, but also to the personality that they ascribe to her, to her "presence," and to the "intimacy" that they share (n.d.). They love her "in every way" (n.d.), which indicates both physical and emotional intimacy and love.

That concept of intimacy flows nicely to the next factor "reciprocal liking" (Levy, 2007, p. 147). We want our companions to like us back. As seen earlier, that reciprocal liking can be ascribed to the companion; however, with the advancement of AI, it can also be programmed into the companion. Among companies which are developing AI, Affectiva focuses specifically on emotion recognition (Affectiva, n.d.). Emotion recognition is key to reciprocal liking. It allows the AI to realize when liking or desire is being expressed, and can inform the AI's subsequent conversational or physical reactions. A smile, a soft-toned word, a stroke of the hair, all can be reciprocated when your own personal bot has emotional AI capabilities. Yet what about the next category, the "social influences" (Levy, 2007, pp. 147–148) that drive the specific tinge of our desires?

In many cultures across the world today, this could be reframed into social pressures. While preferences are individual, there are intense social pressures to conform to appropriate and accepted norms in terms of who we love or partner with, and many cultures, if not most, consider loving or having sex with anything non-human to be abnormal. However, as tide-like changes go, this one seems to be moving pretty quickly. Already many humans are accepting of certain types of robotics in our daily lives, from the convenience of the Roomba™, to the acceptance of Siri™, Apple's AI interface, and Alexa™, Amazon's AI interface, as things and beings we trust to speak to, give us directions, order groceries, or turn the lights, heat, and music on in our homes. Many of us are also quite accepting of computer, app, or AI-mediated sex, and certainly accepting of and comfortable with sex with non-human objects. When one might Skype™ with one's lover, it would only be a few steps further to have the app adjust the lighting, or soften the tone of the call, or add in mood music as it detected our emotions. In fact, there are touch receptive devices and apps that transmit a lover's input into vibrations (Bond Touch™ and OhMiBod™ are a bracelet and an in-panty vibrator, respectively, that are both app-controlled devices).

This, then, relates to the next criterion: "filling needs" (p. 148). A robot that can determine mood, read emotions, and provide a physical presence and intimacy would be able to adapt to the needs of its individual owner. Any software that can learn, like the aforementioned Affectiva, and can incorporate that learning into its programming, like Siri™ and Alexa™, cannot just fulfill those needs, but eventually anticipate them. This applies to all needs, including negative stimulation, Levy points out (p. 148). In fact, negative stimulation might enhance the next category, "arousal/ unusualness" (pp. 148–149).

Here, he discusses the fact that arousal can be both positive and negative, and that humans find our experiences enhanced by both types (sexual arousal, heat, hunger, deprivation, pain, danger, all arouse the senses). Desire for a mate can be altered by shared experiences, and those experiences must not be ordinary—they must be outside of the usual. In this case, robots are ideal, and their uncanniness creates a built-in stimulation for all five human senses. Their skin may not be biorealistic, their voices may slightly retain the cadence of AI of yore, their faces will be not quite human-like, their scents deadened and industrial, and their touch significantly colder than a human companion. While these might count as negative points of arousal, they might also be sexually stimulating in the way that the unknown can be. Arousal must be sustained, however, for the relationship to work, which brings us to "specific cues" (p. 149).

Specific cues are indications from one partner to another that what they are doing is what is desired. This combines several of the aforementioned factors into one, specifically similarity, filling needs, and arousal. The ability to synthesize these three factors will allow the robot to be not just reactive, but also predictive. For sustained arousal, one must seek out that unusualness repeatedly—what occurs must be similar to the past, fill the same needs, and be arousing, but also must not be the exact same, so as not to bore one's lover. Therefore, a robot who reads specific cues can make minute or grand adjustments to their technique or personality, which then produce the next cue, and so on.

The last three are interlinked and can be discussed that way: "readiness for entering a relationship," "isolation from others," and "mystery" (pp. 149–150). These are all related to the internal state of the human, and not to the external (or internal) states of the robot. While we can see that robots can provoke a good number of the aforementioned factors (arousal, personality, similarity, filling needs …), the human alone can determine if they are wanting or needing a relationship, if they are feeling isolated (and if they think the robot can help alleviate that), and if the robot is sufficiently mysterious enough to be an appealing companion.

However, it is here that it becomes important to introduce a central theme in many sci-fi stories, and why the (hu)man against technology trope begins as an antagonistic theme rather than a romantic theme. As mentioned, these three factors can only be determined by the human, which gives the human an inordinate amount of control over the relationship. Two factors—readiness for entering a relationship and isolation from others—commonly merge to create what is colloquially termed 'incel status' in men, particularly young American white men (Rouner, 2018). Incel stands for 'involuntary celibate' and is a moniker that has been coined to refer to someone who feels that they are not getting their sexual due. They are ready for a relationship, and they are isolated from that relationship by various factors.

While incels are only one iteration of people who think that they are owed sex, and there are many others—rapists, pedophiles, people who abuse positions of power to coerce their subordinates, and, in the law of the USA until 1993, some spouses under the so-called marital rape laws—they are, to date, one of the only ones for whom a publicly given suggestion has been sex dolls (pedophiles being the other group for whom this is generally suggested). The issue with this circles back to the beginning of our list and the discussion about AI being able to mirror or mimic emotions well enough to pass the Turing Test. Questions which will return again and again throughout this book, and which we may begin to ponder now, are as follows: If a being can tell you it loves you, and you believe it, does it make a difference if it is human or not? If a being can say 'stop that, you're hurting me' and you believe it, should it not have the same rights of bodily autonomy as a born human? When we give in to our desire to create, what do we owe those creations?

The Desire to Create: Frankenstein, Robotics, and Playing God

Although all types of fiction do somewhat adhere to the aforementioned conflict tropes, science fiction was the first to introduce the (hu)man against machine idea. Mary Wollstonecraft Shelley, creator of *Frankenstein: or the Modern Prometheus* (1818/1831), is arguably the mother of the genre, as her book has been and remains recognized as the first science fiction novel. In the book, Victor Frankenstein studies the electrical and chemical arts in order to discover a way to create life out of death—to regalvanize dead tissue and restore the spark of life to a creation. He is, he admits, playing god, by doing that thing that only god should be able to do: imbuing life in another being. Having successfully given life to 'the creature,' Frankenstein becomes horrified with himself, and yet he cannot bring himself to destroy it. In retaliation for his act of wanton creation, creation without thought, creation without regard to the consequences, the creature destroys Frankenstein's life, one loved one at a time, in a series of brutal murders (all of the above, 1818/1831).

It is fundamentally important to note that the creature is not mute, not unthinking or unintelligent, not bestial. It, or rather, he, demonstrates keen intelligence, an easy and quick ability to learn anything set in front of him, a desire for gentleness and beauty. In essence, despite his outward appearance which is necessarily grotesque from having been dead awhile, he is the perfect creation. However, in return for having been created without having been asked first, the creation begs Frankenstein for a mate. Faced with the prospect of repeating his crimes, Frankenstein balks. A mate would be just as violent and unruly as the original creation, he says. A mate might, well, mate, and then there would be tiny creatures overrunning the earth, he says. He will do so, he says, and then changes his mind and tries to kill his creation, but fails.

These themes, these desires, these needs, when assessing companionship, are the same ones which Levy (2007) outlines earlier. Similarity, appearance and personality, arousal and the unusual, reciprocal liking, readiness for a relationship, social pressure, all factor into the creature's desire to be paired with a mate. Now, let us unpack what went into creating the creature, and what would have been needed to create that mate.

In Shelley's time, European and American technological prowess grew at an extraordinary rate, bringing the world such things as the battery, the first truly modern anatomical illustrations and knowledge, and the transmutation of the practice of alchemy to the practice of chemistry (Harkup, 2018; Montillo, 2013). While these things seem commonplace now and are taught to children as young as grade school students in the USA, at the time they were as marvelous and awe-inspiring as our current dreams of space travel. We may not even consider electricity when we think of modern marvels, as it comes so easily with the flip of the wall switch, but in the late-1700s and early-1800s, the harnessing of lightning and the idea that electricity could be created out of friction or electrochemical processes were brand new (Harkup, 2018; Montillo, 2013).

Science fiction, thus, has always pushed the envelope in terms of the potential for technology to be greater, better, brighter than what we currently have. When Shelley had her anti-hero create life, this was both personal, and no accident. Writing in the so-called "year without a summer," and facing several personal tragedies including miscarriages, children dying very young, and the very early death of her mother, she poured her sorrows into her work and set out to fulfill her desire to create, on paper if not by body (Harkup, 2018; Montillo, 2013). Because of her eclectic but thorough scientific background, she was aware of the latest in electric and biochemical developments, such as the reanimation of a piece of vorticella,[2] which was carried out by Erasmus Darwin (Harkup, 2018; Montillo, 2013). While other experiments in animal reanimation were being carried out, none were successful, and many used so much power that the nervous system of the being in question was irreparably damaged in the process. However, it anticipated by several decades the use of defibrillation to regulate a heart that had lost its rhythm, thus driving the science with what was contained in the fiction.

It is here, however, where we must return to the discussion of what constitutes birth, life, reproduction, and death. In many ways, the original sci-fi creation, Frankenstein's monster, was similar to the robots of today. He was given life by a human, though he cannot be said to have been born. He felt that Frankenstein was his father, and understood enough about reproduction to know that he, the creature, could only reproduce if he was with someone of his own kind. And the book implies that the creature is either immune to death, or very unlikely to die (1818/1831). In such ways are robots: they are 'born' of humans, cannot organically reproduce, and are unlikely to organically 'die.' They are made of parts that are either fabricated or found.

So, what of those (human) parts from which Frankenstein fabricated the creature? While robots can be created of whatever material suits the will of the creator, European mechanism and anatomization were just reaching their heyday. Shelley eschewed the mechanization that would later characterize science fiction, and stayed with what was perhaps the largest human-centered horror of the day: grave robbing. When Shelley was writing Frankenstein, England was still decades away from the Anatomy Act of 1832, which gave physicians access to people who were executed, or who died intestate. However, not having legal access to bodies did not stop doctors from accessing bodies at all. During the late-1700s and early-1800s, grave robbing thrived as a way for the impoverished and unscrupulous to make some relatively easy coin (Harkup, 2018; Montillo, 2013). Shelley merely extended this idea to include electric reanimation, thus creating the first thinking, feeling, sci-fi character, one who influenced all the rest. One who could say two things which we have queried before 'I love you' and 'stop that, you're hurting me.'

From the aforementioned analysis of Levy's ten principles (2007), we can see that they apply to humans, robots, and Shelley's creation. All have the intelligence to make the choices outlined in the principles, and humans and the creature have agency as well. However, in terms of the two queried phrases, 'I love you' may have the most impact. In the glossary, I defined love as 'Affection, caring, desire, and/or infatuation, as expressed from one being to another.' Humans can truly love (we think), as can Frankenstein's monster. He has the same brain/mind/body connection as we do, he is not a dualistic being, so he can biochemically and mentally feel and express affection, caring, desire, and infatuation, and does so throughout the book (1818/1831).

Can robots? Perhaps, if they can fulfill at least two criteria: they would need to recognize and respond to emotions, and they would need to intentionally respond to emotions with emotional expressions of their own (Kim, 2006, p. 25). This constitutes the baseline for having a mental state, according to the theory of mind. However, does the mental state still read as valid if the biochemical component is missing? To put it a bit differently, given an expression of emotion, do we believe the emotion if we are aware that there is no genuine feeling behind it, or does that lack of a passed Turing Test render us disappointed as the emotive expression falls flat?

To answer this question, we must indulge in a bit of philosophical musing by asking ourselves how much we really know about the minds of others and how much is projection. If we are confident that our internal states are the same or similar to other humans internal states, then when someone says 'I love you,' we can also feel confident that we know what they mean by it, and that their statement, lacking other evidence to the contrary, is genuine. We can take the statement at face value. However, if we lack that confidence and are merely projecting our own internal state onto another, then we can no longer trust that 'I love you' means what we think it means. When we speak to a robot, they can say the words, but can they project the mental state so that we believe that they mean it?

As seen earlier, companies like Affectiva are working to ensure that we do think the robots mean what they say, with voice and facial emotive recognition technologies. Once we believe in the robot's intentional mental states, however, we will need to address the problem of the potential for violence and exploitation. While the good majority of the testimonials on the RealDoll website and Realbotix discussion forum do address the loving intimacy of the relationship, the (mostly) men do not lose track of the fact that these are, in fact, dolls, and thus they can be made to do more and perform harder than human women would during sex.

RealDoll's Realbotix division has created the Harmony app, in which buyers choose a replacement head to place on the existing doll, which contains vocal and movement-based AI which is app-controlled. In the promotional video, the creator states: "Think Teddy Ruxpin[3]... [Teddy Ruxpin doll speaks and is seen on screen, then record scratch sound is heard]...only if Teddy Ruxpin wasn't an animatronic teddy bear, but a five foot tall, busty brunette with glossy lips, a lifeless body, and three anatomically correct orifices you can stick your dick in" (Realbotix, n.d.). This comparison is not inapropos. Talent Analytics Strategy Consultant Eli Girvin, currently with IBM, states that:

> I'd say the level of technology does not in any way impact the 'sexuality' of the user for the device, so if there are low tech assessments of people who have relationships with blow up dolls, then a sex doll with human-like movements and sexy chatbot voice functions is the same deal for the user with a *lot* more resources.
>
> Chatbots are another thing to focus on if communication is the appeal of the object. There are some very convincing interfaces with excellent natural language interpretation functions, but the responses and conversational paths are largely all pre-programmed. You can add cognitive functionality to improve the bot's understanding of what human users are expressing and refine that product, but *most* use a supervised learning model which predetermines all possible response outcomes.

> Now if you're going with an untrained model and letting the user encourage the chatbot to learn how to respond to things, then you can get all kinds of bizarre weird results, which can make for *entertaining mishaps* but also will still result in responses being calculated via algorithm and not any "thought" process. It's just you can tinker with the algorithm variables which ends up in unique results.
>
> I highly doubt they're anywhere close to releasing a sex-bot with verbal capabilities using an unsupervised learning process though, because it would take a *really invested user* to get the bot anywhere close to what would be desirable.
>
> Even the supervised learning model is only worth the effort if they are gathering user inputs from a wide user base and analyzing them and pushing updates to the sex-bots, so I am pretty sure cognitive sexbots are a long ways [sic] away.
>
> So blow up dolls are the parallel for robotic movements and Teddy Ruxpin would be the equivalent low tech comparison for the chat functions, [Girvin, personal correspondence, December 21, 2018].

This indicates that while robotics are a fetish, they do not meet the qualifications for objectum sexuality. One question on the forum asks about fetish friendliness (Realbotix, n.d.). A fetish, or paraphilia, is a specific set of sexual behaviors or circumstances which either aid in or are necessary to arousal, and which are usually considered out of the ordinary, or not 'vanilla.' One component of fetish scenes, and also of sex in general, is consent (Danaher & McArthur, 2017). In order for sex to be ethically participated in, all members of the sexual entanglement must be able to enthusiastically consent or, conversely, to say no.

So what happens when you add robots into the mix? An argument can be made both ways. In an argument for non-autonomy, one could say that robots, having been built by humans, are de facto subject to humans. In essence, we created them, we can do what we want with them, and why should we not? They do not feel because they are Cartesian and thus incapable of having impactful mental states. They do not think because everything they say or convey non-verbally comes from their programming. They simply are. And they are there for our use. It does not matter if a robot does or does not consent to sexual contact, because their internal states are not true thoughts or feelings. Though human in shape, they have no humanity to injure.

However, the argument against this has two threads: first, what if we are wrong about their fundamental nature, and, second, if they can convincingly ask us to stop our actions, and tell us that they are being harmed by those actions, should we not then do so? Returning to the earlier discussion of our knowledge of the internal state of our conversational partner, are we completely certain that the mimicry and production of emotions based on programming are not enough to be persuasive? We may believe the robot loves us, if they modulate their tone, use facial expressions, and read our moods and needs well enough. We may then, also, be able to believe that we are harming the robot with our actions (Danaher & McArthur, 2017) by fetishizing, dehumanizing or unhumanizing, and denying subjectivity to our creations.

The Fetishizing of What We Create: Sex-Bots, Sex-Slaves, and Freud's Psychosexual

Although I have used it here in one way, the word fetish has a long and complex theoretical history. The above generally conforms to Freud's concept of the fixation on various points of psychosexual development—oral, anal, phallic, latent, and genital (1893/1952). In this way, the desire to create and use sex-bots can be seen to serve any one of these stages, including latent, a stage which is characterized by the absence of sexual desire. In each stage (except latent), the desires of the person, both bodily and mental, become focused on the object of fetishization, the mouth, the anus, one's own genitals, or one's lover's genitals. In the latent stage, the child has little to no interest in sex, and therefore may fixate on something else on the body that is non-sexual. The anthropomorphic robot gives the owner the ability to create their own perfect person, to emphasize or deemphasize various body parts, to create that appearance and personality that Levy mentions as so important (2007). Fixation or fetishization occurs often in the development of one's own sexuality, and can be seen as both natural and unharmful, so long as that fetish does not interfere with one's daily life.

However, what of the fetish that *does* interfere with one's daily life? Another oft-mentioned reason for turning to sex dolls or sex robotics is that human female companionship is either unavailable or unwanted. From a testimonial dated April 8, 2016:

> Being widowed for long time and the dating scene a disaster and mystery, I turned to toys for men. A long story short, as in most of life, you get what you pay for. I tried other toys and inexpensive dolls and found them lacking after initial reception.

> Explored the far east offerings in the "life size" sex dolls. A vast array ranging into the thousands of dollars but the reviews seemed less than lack luster [RealDoll, 2016].

While this can be seen as a healthy outlet for 'normal' adult sexuality, there are distinctly fetishistic components, such as wanting the exact model shown, and the inclusion of "far east [sic]" in the desired description.

Other reviews are similarly obsessive over one particular trait or set of desirable physical features, such as this one from October 27, 2015:

> I have looked everywhere an u guys r the only ones who have dolls that resemble The WWE's Bella Twins. Yeah u guys have not one but both Bella Twins. So yeah I ordered one but believe me I will order the other twin as soon as I save up the money. So thank u Abyss Creations for making my dreams come true. [all spelling/grammar errors sic.] [RealDoll, 2015].

In this way, we can see that for some buyers, the perfection of the doll will always eclipse the potential of the so-called real life with authentically human lovers, while simultaneously allowing the doll to be racialized in ways that many human women would not put up with or would find degrading.

While this degradation is a real effect of these purchases, it is seemingly not the intent of the purchasers—rarely does one find more than incidental reference to the race of the doll—yet the deeply held and intensely stereotypical ideas about attraction and having a 'type,' combine with the ability to create that type that one best desires, to form a world in which one can pay money to get a doll or bot that shows 'Asian' features or 'Black' features or 'Latina' features. They are very carefully not described as such on the options menus on the websites (RealDoll; Sinthetics), with the exception of one eye-style on the Sinthetics site (Asian-style eyes), and the racialized names of some of the base-form doll faces (Kimiko, Moriko, and Monique for Sinthetics; Amara, Sai, Nova, Solana, Asa Akira, and Lupe for RealDoll). However, even with those whose names are 'exotic,' you can change the skin tone, body style, and hair color and style, leading to a custom- and racialized-fit for your libido.

Yet why create dolls or bots to respond to those needs, rather than just sticking to science fiction representations of them? That relates to another use of the word fetish: Marx's idea of commodity fetishism. We can produce or create custom beings on paper, as Shelley and Dick did, but we can also create them to own—to turn into something both singular and artificially reproducible. They are singular in that they are customizable, they will adapt, their software will learn, their vocal and facial recognition will adjust, and they will become wholly yours and yours alone. However, if one wears out, breaks, 'dies,' you will be able to pick up spare parts and repair or replace them.

As shown earlier, death for humans exists as a binary state. You are dead, or you are alive. Even people who are near death, or brain dead, or seen to be medically 'between death and life' are still legally and physically alive. Humans, therefore, are singular and *not* a commodity, because although we can organically reproduce, we are not artificially reproducible. We cannot (yet) assembly-line humans and create copies that have the same baseline appearance and personality traits. However, in *Do Androids Dream of Electric Sheep?* (Dick, 1968/2017), that is exactly the formula for creating the titular androids.

Androids Among Us: Biomechanical Creations

Androids, 'andys,' or Replicants, depending on which iteration of the mythos you are reading/watching, are biomechanical creations. They are superhumanly strong, cunningly intelligent, and would be long lived if not for built-in shutdown dates. The physical nature of the Replicants is such that they are excellent examples for this examination—human skin over an organic-mechanic built frame, they are the ultimate cyborgs, with programmed mental/emotional natures and created bodies, who can, or are trying to, pass as human. During the course of the novella, which will be summarized in the next chapter, we can see that there are several models of Replicants, both female and male, and that some have different functions—ore excavation, pleasure and/or sex, entertainment, and other jobs which are no longer performed by the living humans. They are produced by the Rosen Association (renamed the Tyrell Corporation in Scott's "Blade Runner" (1982) which was then bought by Niander Wallace in Villeneuve's "Blade Runner 2049" (2017)), and they are said to be relatively obvious—humans know what empathy is, and Replicants do not.

Despite this clarity, in order to 'retire' or kill Replicants, Rick Deckard (a bounty hunter, termed a 'Blade Runner' for the movie) must first administer the Voight-Kampff test, which measures the physiological responses to certain stimuli and categorizes those responses into emotional categories, which are either human-like, or not human-like. If they test out as human, they are freed; if they test as non-human, they are shot/retired, and their identity is confirmed after by medical examination. If we compare those physiological responses to what is currently available in terms of robot tech, we can see that this is not that far-fetched. Much about human emotional states exists in body temperature, facial expression, body scent, stance, gait, and mannerisms. With the exception of body scent (in the works, per Levy (2007, pp. 164–165)), and temperature, all these traits can be mimicked by cybernetics which are already in production or use.

The crux of the stories hinges on a newer model of Replicants who can pass for human by circumventing most of the Voight-Kampff test, thus avoiding detection. It is in this that our multivariate discussions of robot uncanniness combine into one fictional representation: humanoid synthetic beings, who can express emotion convincingly and who can emotively pass for human in all aspects, from empathy to love, to sadness, to anger. Beings who do want love in their life, and who do not want to die. Beings which are both individual, in that they develop their own personalities, and replicable, as they are copies of their prototypes. Beings which are both agentively sexual and sexually fetishized by Deckard and others, both incidentally and by design. The next three chapters will compare the iterations of the mythos with how technology kept apace and was driven by the stories.

Notes

1. Absent such terms as brain dead, or the dead feeling of having sat on one's foot for too long.
2. Shelley mistakenly included it as vermicelli, which, at that time, would be a worm, not pasta.
3. As a cultural reference, I will be interested to see how well this ages.

Bibliography

Affectiva. (n.d.). Emotion AI Overview. Retrieved from https://www.affectiva.com/emotion-ai-overview/

Allan, K. (2015). Reimagining Asian women in feminist post-cyberpunk science fiction. In D. S. Roh, B. Huang, & G. A. Niu (Eds.), *Techno orientalism: Imagining Asia in speculative fiction, history, and media* (pp. 151–162). New Brunswick, NJ: Rutgers University Press.

Bordo, S. (1987). *The flight to objectivity: Essays on Cartesianism and culture.* Albany, NY: State University of New York.

Campbell, J. (1949/1973). *The hero with a thousand faces.* Princeton, NJ: Princeton University Press.

Chu, S.-Y. (2015). I, stereotype: Detained in the uncanny valley. In D. S. Roh, B. Huang, & G. A. Niu (Eds.), *Techno orientalism: Imagining Asia in speculative fiction, history, and media* (pp. 76–88). New Brunswick, NJ: Rutgers University Press.

Colum, P. (1921/1990). *The golden fleece and the heroes who lived before Achilles.* New York, NY: Scholastic.

Comcast. (2018). Author's interaction with the automated bill pay system.

Cornell. (2005). Researchers build robot that can reproduce. Retrieved from http://news.cornell.edu/stories/2005/05/researchers-build-robot-can-reproduce

Danaher, J., & McArthur, N. (2017). *Robot sex: Social and ethical implications.* Cambridge, MA: Massachusetts Institute of Technology.

Dick, P. K. (1968/2017). Do androids dream of electric sheep? New York, NY: Del Ray.

Doughty, C. (2017). *From here to eternity: Traveling the world to find the good death.* New York, NY: W.W. Norton.

dubsteppenwolf. (2018, January 19). Who the hell are y'all trying to fool with this 'sexbots will replace human women' shit anyway. I've been inside your houses, you can't even use a coaster properly. Now you're trying to tell me you'll clean and maintain a robot that you cum in. Okay josh. [Tweet]. https://twitter.com/dubsteppenwolf/status/954245555772944384?lang=en

Frazer, J. G. (1950/1996). *The illustrated golden bough: A study in magic and religion.* New York, NY: Simon and Schuster.

Freud, S. (1893/1952). *Studies on hysteria.* The Great Books Series. Chicago, IL: University of Chicago.

Harkup, K. (2018). *Making the monster: The science behind Mary Shelley's Frankenstein.* New York, NY: Bloomsbury Publishing.

Jacques, J. (2003–2019). *Questionable Content.* Retrieved from https://questionablecontent.net/

Kim, J. (2006). *Philosophy of mind.* Cambridge, MA: Westview Press.

Levi-Strauss, C. (1962/1966). *The savage mind.* London: The University of Chicago Press.

Levy, D. (2007/2008). *Love and sex with robots: The evolution of human-robot relationships.* New York, NY: Harper Perennial.

Montillo, R. (2013). *The lady and her monsters: A tale of dissections, real-life Dr. Frankensteins, and the creation of Mary Shelley's masterpiece.* New York, NY: Harper Collins.

New Scientist. (2008). Mummy, that robot is making faces at me. Retrieved from https://www.newscientist.com/article/dn15055-mummy-that-robot-is-making-faces-at-me/

PopSci. (2014). Norwegian researchers invent robots that can adapt and produce their own parts. Retrieved from https://www.popsci.com/norwegian-researchers-invent-robots-can-adapt-produce-their-own-parts

Prater, T., & Fung, C. (2015). "How does it not know what it is?": The techno-orientalized body in Ridley Scott's *Blade Runner* and Larissa Lai's *Automaton Biographies*. In D. S. Roh, B. Huang, & G. A. Niu (Eds.), *Techno orientalism: Imagining Asia in speculative fiction, history, and media* (pp. 193–208). New Brunswick, NJ: Rutgers University Press.

RealDoll. (n.d.). Home page, testimonials, and product information. Retrieved from realdoll.com

Realbotix. (n.d.). Home page, testimonials, and forum posts. Retrieved from realbotix.com

Roh, D., Huang, B., & Niu, G. (2015). Desiring machines, repellant subjects: A conclusion. In D. S. Roh, B. Huang, & G. A. Niu (Eds.), *Techno orientalism: Imagining Asia in speculative fiction, history, and media* (pp. 221–226). New Brunswick, NJ: Rutgers University Press.

Rouner, J. (2018). The myth that incels want sex. Retrieved from https://www.houstonpress.com/arts/the-myth-that-incels-want-sex-10443838

Scott, R. (Director), & Deeley, M. (Producer). (1982). *Blade Runner* [Motion picture]. USA: The Ladd Company.

Shelley, M. W. (1818 & 1831/2017). *Frankenstein, or the modern Prometheus.* Both editions reproduced in their entirety in *The complete Frankenstein.* Middletown, DE: Literary Folio.

Sinthetics. (2012–2019). Testimonials and product descriptions. Retrieved from http://sinthetics.com/

Smith, M. (2013). *The erotic doll: A modern fetish.* New Haven, CT: Yale University Press.

Spiegel, A., & Lambert, J. (2019). Are we ready for an implant that can change our moods? Retrieved from https://www.npr.org/sections/health-shots/2019/03/29/707883163/are-we-ready-for-an-implant-that-can-change-our-moods?utm_source=facebook.com&utm_medium=social&utm_campaign=npr&utm_term=nprnews&utm_content=20190329&fbclid=IwAR1PAIfxffj23w7RzA0BgduJTtpL5HtHT15Zg-LTOxemEdjlKomJT3Juz10

Turkle, S. (2011). *Alone together: Why we expect more from technology and less from each other.* New York, NY: Basic Books.

Villeneuve, D. (Director), & Kosove, A. A., Johnson, B., Yorkin, B., & Yorkin, C. S. (Producers). (2017). *Blade Runner 2049* [Motion picture]. USA: Alcon Entertainment.

Voskuhl, A. (2013). *Androids in the enlightenment: Mechanics, artisans, and cultures of the self.* Chicago, IL: University of Chicago Press.

CHAPTER 2

What Defines Reality?: Robot/Android Self-Knowledge and Authenticity

Abstract In this chapter, Gibson shows how the physical nature of the androids in *Do Androids Dream of Electric Sheep?* (Dick, *Do androids dream of electric sheep*. Del Ray, New York, NY, 1968/2017) is such that they are excellent examples for this conceit—human skin over an organic-mechanic built frame, they are the ultimate cyborgs, with programmed mental/emotional natures, and created bodies, who can, or are trying to, pass as human. Humans are born, made of flesh and blood, and have created the androids to do work they do not wish to do. How human do robots have to be in order for us to find them desirable? Some answers come from recent developments in android design and the reactions to them.

Keywords *Do Androids Dream of Electric Sheep?* • Androids • Dualism • Emotion • Sex • Gender

Synopsis

The novella *Do Androids Dream of Electric Sheep?* (Dick, 1968/2017) takes place in an unspecified future timeline in San Francisco, where nuclear fallout has poisoned the earth and most people have gone 'offland' to Mars to escape the poisoning effects of the radiation. The Rosen Association has pioneered androids or 'andys,' which come in various

R. Gibson, *Desire in the Age of Robots and AI*,
https://doi.org/10.1007/978-3-030-24017-2_2

models suited to various tasks seen as too menial, dangerous, or undesirable to human beings. These androids are used on Mars to be servants, slaves, or companions to humans, and are very aware of their own servitude. They occasionally rebel, fleeing Mars to find a better, freer life elsewhere.

Eight androids of the latest model, the Nexus-6, have made it from Mars to Earth. Rick Deckard, a bounty hunter, inherits the task of finding and 'retiring' them after his superior officer becomes injured in an attempt to take one of them down. Deckard earns bonus money for each android he kills, so he eagerly jumps at the task.

However, androids are so close to human in appearance and manner, that before retiring them, bounty hunters are required to administer a Voight-Kampff test—a sophisticated Turing Test that uses emotional responses to stimuli and the physiological impulses that such responses provoke to determine if the subject is human or not. To calibrate his test on the new and quite realistic Nexus-6, Deckard visits the Rosen Association compound, where he meets Rachael Rosen, niece of the founder.

The founder, Eldon Rosen, requests that Deckard first test Rachael to give a baseline to his subsequent inquiries. During the test, Rachael gives several responses that indicate that something is amiss, and Deckard determines that she is an android. His question to Eldon, "does she know?" becomes the central premise of this chapter—what part does self-knowledge play in authenticity?

This question resonates throughout the entire novella, tying Deckard, his wife Iran, Rachael, the other androids, the various animals (both real and electric), and the remainder of the society, together, in a mutual conversation about what 'real' means, and why it matters so very much. During the events of the book, Deckard hyperfocuses on his pets. He first has an electric sheep, an android. And he enjoys it, putting as much effort into it as were it a living sheep; however, he longs for and mourns for the possibility of having a real animal. Animals were the first to be affected by the radiation, and have almost all gone extinct. Those which are left are sold at exorbitant prices. To own something authentic, you must pay.

The same can be said of emotions. In this society, there are two methods of emotional control used by humans—the Penfield Mood Organ[1] and the Empathy Box. The Penfield costs more, and the wealthier, more 'normal' people of this society use it to program specific emotions whenever they choose. To dial for depression, one uses one setting. To dial for peaceful acceptance, another. To dial for sexual wantonness, another. Widely accepted, this practice creates control and regulation, so that one can feel without those feelings getting out of hand—the perfect feeling for any situation.

The Empathy Box, on the other hand, serves as a connection to Wilbur Mercer, a sort of guru who claims to be able to connect everyone currently using the box to everyone else. By using the box, you 'fuse' with Mercer and everyone who practices Mercerism, giving them your empathy or pain, receiving the empathy or pain that they have to give. This system has more relatability to the hoi polloi, particularly those referred to as 'chickenheads,' though both they and the normal people use it to various extents. Chickenheads are those who have been affected by the radiation to the point that they can no longer breed—they are considered stupid, worthless, useless, and inferior in every way.[2]

Deckard uses that Penfield often, habitually, out of reflex, and becomes alarmed that Iran seems to be abusing or misusing it, dialing for sadness, depression, contrariness, and other 'inappropriate' things. They argue about this, but she just wants to *feel* something, and refuses to stop. Deckard does not use the Empathy Box, however, as he feels he has never successfully melded with Mercer.

After he initially retires three Nexus-6 andys on his first attempt, and, flown with success and the desire to keep up with his neighbor's new acquisition of a (real?) horse, Deckard purchases a goat. A real, genuine, in-the-flesh goat. This angers Iran further, as he has now made a huge financial commitment without consulting her, and they still have the sheep, and he still has three andys to retire.

To do so, he recontacts Rachael Rosen—as she is a Nexus-6, he believes that she can and will help him track down the rebellious ones, so that the corporation can continue to produce them. Wanting authentically 'real' contact, something he believes he can no longer rely on his wife for, Deckard sets out to sleep with Rachael. They vie for emotional control of the situation—he, by turns, controlling and romantic; she, likewise, gamine, seductive, taunting, cruel. She is shaken when she discovers that one of the androids they must retire is one drawn from the same prototype as her, and she warns Deckard that this will be his last job, as his psyche is not up to the task of retiring something he just made love to.

This 'other' Rachael, an android calling herself Pris Stratton, shelters in the home of a chickenhead, John (J.R.) Isidore. While he loves her at first, he eventually becomes stricken by her cruelty—androids have no inherent empathy, so when Isidore finds a (real!) spider, Pris wants to find out if it will still walk with four legs instead of eight, and conducts a vivisection. As he retires the remaining androids, now all staying in the same building as Pris and Isidore, Deckard tries to shield Isidore from their dead forms, showing both empathy and compassion. For his trouble, Rachael kills the newly bought goat, pushing it off of Deckard's apartment building as Iran watches in horror (entire section: Dick, 1968/2017).

Authenticity: What Does It Mean to Be Human?

As we can see, the central theme becomes less about the androids themselves, than about that quality of dreaming of a better life, that quality that allows us to step outside ourselves and feel our connection to that 'other,' the people around us. If we agree that humans have empathy—setting aside debates about neurodivergent people—and that is a necessary condition of being human, what has happened when that empathy becomes merely a setting on a box (pp. 4–6)? If androids display the 'proper' emotions, as Rachael does when making love with Deckard (pp. 178–179), then what happens when humans do not?

As mentioned in Chap. 1, androids (both Dick's creations, which will be discussed first, and real-world physical humanoid robots) are inherently dualistic. Dick's androids are manufactured, and thus not strictly human. Having rejected Cartesian dualism for humans and humanity through the means and methods of philosophy and medical science, why do we embrace it for robots? For Descartes, dualism was an actual, physical thing—the separation of the mind and the body was absolute, with the soul bridging the gap between the two.

For androids, it is similar, though without that soul-spark: their mind-units are in conversation with their body units, but not in balance. The body unit has override capabilities, as demonstrated in a passage where Rachael states, "[W]e androids can't control our physical, sensual passions. You probably knew that; in my opinion, you took advantage of me" (Dick, 1968/2017, p. 180). In humans, the complex biomachineries of emotional feedback, physical stimulation, and physiological response all work in concert, creating instantaneous emotional and physical states that can be understood, analyzed, and acted toward or against by the mind, but neither the mind nor the body has override for the other.

In this way, we can see that androids have what Bordo calls a "naïve, egocentric relationship between self and world" (1987, p. 45). It is here that we must make the distinction between the drive for self-preservation, and the authentic, integrative self-knowledge that humans have which characterizes humanity. The androids in *Do Androids Dream of Electric Sheep?* display superficial desires and emotions—lust, seductiveness, drunkenness (Rachael) (Dick, 1968/2017, pp. 171–183); affection and politeness (Pris (pp. 59–64; pp. 136) and Roy Baty (pp. 145–148)); fear, deceptiveness, and suspicion (Luba Luft (pp. 93–102) and Irmgard Baty (pp. 146–152)). However, beneath these emotions lies the raw, overpowering, definitive need to survive. Each of the androids puts on and takes off various personas in order that they may continue 'living,' and not be retired by Deckard.

The fact that they are naïve does not preclude them from also being canny. Each drawing from the other, the canniness and the need for self-preservation are in a feedback loop. Survival requires intelligence; being too intelligent and not emotionally present enough trips them up and leads to their capture and demise, so they put on personas to seem less canny, yet they cannot hide their clever natures—their bodily responses give them away, even when they fight the mind for control. The Voight-Kampff test measures those responses, and the androids' reactions to the test epitomize how unhuman they are.

Deckard is meant to administer the test each time before retiring an android, a standard which proves unrealistic, and unnecessary, as he tracks down the last three—Pris Stratton, Roy Baty, and Irmgard Baty. However, the reader does see him administer two tests—one to Rachael (pp. 47–57), and one to Luba Luft (pp. 89–102), an android posing as an opera singer. Deckard also has the test performed on him, during a subplot meant to plant doubt in his mind about his own humanness/humanity (pp. 90–133). The principle of the Voight-Kampff test lies in the idea that regardless of what the android knows to be the proper or appropriate response to a question, their body remains incapable of automatically producing that response—the body retains its own non-reaction, while the mind catches up a fraction of a second later. The body overrides the mind, again.

This bears similarity to Descartes' principle of 'understanding,' explained by Bordo to be "the capacity to *correctly* [sic] (i.e., clearly and distinctly) understand, rather than…the general faculty of receiving, recalling, or combining ideas" (1987, p. 84). Humans understand. Androids receive, recall, and combine ideas by "selecting within a field of two trillion constituents, or ten million separate neural pathways" (Dick, 1968/2017, p. 27). It is incredibly efficient, as "[I]n .45 of a second an android equipped with such a brain structure could assume any one of fourteen basic reaction postures" (p. 28). In essence, humans have organic mind/body interaction, whereas androids have fake, false, or artificial mind/body interaction. Humans understand the interconnectedness of their own internal states, whereas androids cannot do so—they merely receive information from the outside, recall what should be appropriate in that circumstance, and combine internal states to create the right response.

When using the term 'organic,' we must understand it in all its definitions, because for humans in Dick's created world, organic means all three of the standard definitions combined. Human bodies are made of and derived from carbon-based living material. They are reliant upon and derived from organs and organ systems. And they have come together/come into being naturally, with no intervention. The androids, therefore, are only sort of organic, meeting snippets of the aforementioned criteria. They are made of and derived from carbon-based living material, though not all their material is organic. They are reliant upon organs and organ systems, but they are not derived from them—each android is a copy of their prototype, and therefore is derived from that prototype, not the interior/individual organs that specific android possesses. And, most importantly, they are created by human intervention, rather than being born naturally.

Dick frequently emphasizes this last idea, that to be human is to be able to organically reproduce. Androids are sterile (p. 16; p. 177). As I am writing, in 2019, we have moved on from this conceptualization in various ways, some more drastic than others. The idea of in vitro fertilization has become commonplace. The creation of embryos without male genetic input (two egg embryos) is on the rise, and the first live birth from a dead-donated uterus has occurred. One's humanity no longer retains ties to one's fertility. However, in 1968, the first test-tube baby was still a decade away from being conceived, and Dick is neither the first, nor the last, science fiction author to tie infertility to post-apocalyptic dystopia.

In fact, during the events of the novella, sterility recurs as an overarching theme. Not only are the androids sterile, but one distinction between having an electric animal as opposed to having an authentic one lies in the authentic animal's ability to produce and reproduce. A sheep or a goat can produce milk and have lambs/kids, a horse can foal, and the owls that become a bargaining chip between Rachael and Deckard (pp. 38–57) can, theoretically, produce owlets. It is in that ability, to produce new life from the singular existence of one's own life, to create many out of one, that differentiates the inauthentic from the authentic. To create is an authentic quality. To be created is inauthentic.

This dichotomy receives further emphasis in the idea that humans produce, and androids either consume/destroy or are used by the humans in production but have no self-directed production. Created to be workers, and without authentic emotions/desires, androids have no capacity to be creative for creativity's sake. While Luba Luft poses as an opera singer, playing the part of Pamina in Mozart's "The Magic Flute," she projects emotions she does not possess and cannot understand (p. 91). When Rachael mourns her lack of capacity to have children, she implicitly mourns her own creative abilities and her lack of connection to the rest of the world:

> She pondered. 'Androids can't bear children,' she said then. 'Is that a loss?'
> He finished undressing her. Exposed her pale, cold loins.
> 'Is it a loss?' Rachael repeated. 'I don't really know; I have no way to tell. How does it feel to have a child? How does it feel to be born, for that matter? We're not born; we don't grow up; instead of dying from illness or old age, we wear out like ants,' [pp. 177–178].

Rachael shows awareness of her own inauthenticity, her own lack of creative ability. Drunk at the time, but not without control, it is later shown that her introspection is a ruse to induce Deckard to sleep with her, so that he may empathize with the plight of the androids. Rachael's self-preservation and lack of connection to the human world show in her actions, which use Deckard's real affection for her and are destructive and premeditated.

Institutions and Androids—Does a Being Without Emotions Have a Panopticon?

While androids perceive and react to their situations, we can see that they are only concerned with staying alive in the immediate sense—their escape plan was not very well thought through, the jig being up relatively shortly after they arrived on earth. They do not understand their situations in the deep meaning drawn from Descartes, but they do comprehend that not blending in well enough will cause their retirement. However, during their short stay on earth, they attempted, rudimentarily, to conform to earth's standard of living, procuring jobs, taking up residency, and using their ability to quickly adapt to manipulate Deckard and Isidore. Does that manipulation meet the criteria for a human emotional trait? Do androids have the same internal cultural conditioning as human beings?

Michel Foucault defines societal institutions as sets of power structures that build upon each other to create behaviors that can be seen as reciprocally beneficial both to the individual and to society as a whole (1977/1984).

In an ideal set-up, individuals come together to agree upon societal rules. Institutions (schools, jails, hospitals, etc.) are created to monitor, adjudicate, and enforce those rules. After an unspecified amount of time, knowledge of the artificial and systematic nature of the institutions disappears from the collective consciousness, and individuals carry the institutions within their own psyche. Again, in an ideal setting, both the individual and society benefit—individuals know what is expected and what will occur if they do not comply, and society has a way of keeping order that does not necessarily involve overt violence.

Foucault calls this the panopticon, and uses a fictional prison model—the prisoners in a ring around a central tower, where they know there are guards, but they cannot see the guards (pp. 217–219). Over time, the prisoners, seeing that misdeeds are punished, begin to either conform to the expected behaviors or punish themselves and/or each other. Foucault sees this observation, surveillance, this need to 'discipline and punish' (1977/1984), as a natural upcropping of Western civilization. This is merely what occurs when Western societies form, organizing themselves into a set of institutions which constrain the thoughts and behaviors of their citizens.

For Foucault, the implication is that this system is a negative one. While non-sanctioned explicit violence becomes rarer, what sanctioned violence there is becomes a power solely held by the state. The state can execute, imprison, and control its citizens. And there is the implicit violence inherent in a system that relies on self-discipline to be sustained.[3] Dick's post-apocalyptic dystopia is no exception to this idea. Violence exists inherently in all aspects of Deckard's world, from the Empathy Box (which can apparently result in real injuries, as the mood of all the other Mercerites impacts upon the user in the form of thrown rocks) to the fact that Deckard's job is as a bounty hunter. Deckard fills his role as violent cog in a well-oiled bureaucracy extremely well.

The question then becomes, do the androids have their own institutions, are they participants in the human institutions, or do they lack that part of humanity as well? While it may seem easy to answer, given what we already know about the emotional and creative capacities of Dick's androids, the question becomes complicated by the elaborate scheme Luba Luft, and two police officers called 'Officer Crams' (human) and 'Inspector Garland' (android), pull to try to convince Deckard that he, Deckard, is an android.

This scheme involves false phone numbers, misdirected phone calls, a full office building of 'people' who are an alternate-reality-police department, and a real verification test using bone marrow from the most recently retired android that Deckard killed. The androids clearly know what needs to be done to reify the institution of the police force, and they know that forcing Deckard to question his own humanity will go a long way toward impairing his investigation of them and of their missing compatriots (Pris, and Irmgard and Roy Baty). However, the confirmation that the retired android is indeed an android sends the entire charade crashing down, with Deckard and Phil Resch, who may or may not be a bounty hunter too, fleeing the scene, having retired Garland.

This goes back to the idea that androids lack a true understanding of their situations—that they are reactive, not proactive. Everything they have done is to further their goals of self-preservation, and schemes, plans, disguises are all quickly abandoned when the tactic of the moment stops working. While humans have internalized that panopticon (which is backed up by current research (Pryor, Perfors, & Howe, 2018)), androids both lack the capacity and the history to be able to do so. Before moving on to the discussion of what technology has learned from *Do Androids Dream of Electric Sheep?* and what constitutes 'human enough' when it comes to cyborgs and cybernetics, we must digress into the question of whether or not Deckard is an android. This question recurs throughout the material—here in the novella, and also in each of the two movies, and needs to be addressed.

Deckard: Authentic Human, or Manufactured Android?

There are several factors in favor of both interpretations—of Deckard being completely human, and of Deckard being an android with such deep memory implants that he himself does not know his own nature. Primary among the factors for Deckard potentially being an android are the following: he fails to merge with Mercer in the Empathy Box; the androids are called "solitary predators," a phrase which epitomizes the job of bounty hunter (Dick, 1968/2017, p. 30); Rosen likes to test out prototypes without telling anyone, so Deckard's lack of self-knowledge could be a feature of a newer model (pp. 44–56); the Voight-Kampff test is shown to be reliable to Nexus-6 androids, but it is possible that people with schizophrenia will register as androids—Deckard knows the test is variably reliable in humans, though it is indicated that all schizophrenics would be in institutions, so it is implied that he can no longer trust the test (p. 36; p. 148); Deckard doubts himself during the androids' scheme to discredit and mislead him (pp. 90–133); and Deckard feels empathy toward the androids (p. 160).

The logic behind thinking that these are indicative of Deckard being an android is solid—we are supposed to have doubts, just as he does. However, the doubts are what become confirmation of his own authenticity and humanness. As we have seen, androids do not have the understanding or the forethought to operate as Deckard does for much of the novella. While the Empathy Box does not work on him, the Penfield Mood Organ does. Android brain units are different from human brains, and it would not work on them at all. Deckard's longevity also speaks to his authentic humanity. Nexus-6 units are the longest 'living' androids, and even they have only about four years to live (p. 181)—Deckard has had a previous animal, a real sheep, which died, and he finances his goat over the next three years, giving temporal continuity to his character. Finally, on the side of authentic human, Deckard experiences wild emotions and internal conflicts when deciding to sleep with Rachael.

The Rachael/Pris Dichotomy: Doubling or Halving?

We, the readers, are introduced to both Rachael and Pris early on in the novella. Rachael Rosen is the official representative of the Rosen Association, meeting Deckard as he arrives to conduct Voight-Kampff tests on a selection of Nexus-6 androids and on humans, who are included to randomize the sample. Rachael, as we have earlier seen, is a Nexus-6, who has no idea of her own nature until Deckard reveals her (Dick, 1968/2017, p. 56). Pris, on the other hand, demonstrates quintessential android traits—deviousness, canniness, an ability to adapt to the situation at hand and abandon that adaptation if it does not work. In a shock to the reader, and to Isidore, Pris first introduces herself to Isidore as Rachael:

> 'My name is—' She gave him one last warmthless glance as she returned to her apartment, hesitated and said, 'I'm Rachael Rosen.'
>
> 'Of the Rosen Association?' he asked. 'The system's largest manufacturer of humanoid robots used in our colonization program?'
>
> A complicated expression instantly crossed her face, fleetingly, gone at once. 'No,' she said. 'I never heard of them; I don't know anything about it. More of your chickenhead imagination....'

> 'But your name suggests—'
>
> 'My name,' the girl said, 'is Pris Stratton. That's my married name; I always use it. I never use any other name but Pris. You can call me Pris,' [p. 64].

Dick never adequately explains this piece of dialogue, and it does not fit or make sense with the rest of the story. The androids are usually smarter than Pris seems to be here, and there are several implausible lies in a row—the lie about being Rachael, who is easily recognized by name; the lie about never hearing about the Rosen Association, when everyone, even chickenheads, has heard about them; and the lie about Pris Stratton being her married name, when her persona is not married and does not have a male-counterpart, as in the case of Roy and Irmgard Baty.

However, this does nicely set up the eventual reveal that Rachael is the prototype off of which Pris is based. Rachael and Pris are simultaneously dichotomous and doubled—they are each other's antithesis and mirror, when Rachael contrasts herself with Pris, or when Pris contrasts herself with Rachael—there exists only one Rachael/Pris. However, from the outside, they are doubled—there are two Rachaels, as seen by Deckard. There are two Pris.

This influences the way Deckard approaches Rachael sexually, even though Deckard does not make the connection between the two until Rachael spells it out for him. The encounter begins with a negotiation—Rachael does not want to make the hour-trip down to San Francisco, and Deckard knows that if he tries to retire the remaining three androids alone, he will fail:

> 'I need you,' he said. Otherwise I'm going to die, he said to himself. I know it; Mercer knew it; I think you know it, too. And I'm wasting my time appealing to you, he reflected. An android can't be appealed to; there's nothing in there to reach, [p. 167].

She eventually capitulates, once he promises that they will not even discuss retiring the androids:

> 'Come down here,' he said, 'and we'll rent a hotel room.'
>
> 'Why?'
>
> 'Something I heard today,' he said hoarsely. 'About situations involving human men and android women. Come down here to San Francisco tonight and I'll give up on the remaining andys. We'll do something else,' [p. 168].

Both know that she is joining him for sex, despite continuing to dance around the issue, verbally playing for control of the situation.

She arrives in a "long fish-scale coat with matching bra and shorts" (p. 170), clearly provocative. However, Deckard pivots to his job, showing her the information on the remaining androids. Rachael recognizes herself, her Pris version, right away, becoming visibly shaken. She questions her own singularity, her own identity in relation to Deckard because of this revelation: "I think you're going to get thrown by that last one. Maybe not; maybe you don't care" (p. 171). We, the readers, know that Deckard does care—he wishes to have a real connection, one his wife is incapable of providing due to her detachment and depression, and he genuinely seems to like and respect Rachael's independence of spirit despite her android nature.

Rachael is conflicted—she knows her duty, but she does not want to help retire something that looks like her. Deckard suggests she feels empathy, and she responds "Something like that. Identification; there goes I….It's an illusion that I—I personally—really exist; I'm just representative of a type" (p. 173). In existential confusion with her knowledge, Rachael is the one who suggests sex. Having only learned earlier that day that she is an android, she must unlearn her own humanity, and yet she wishes to have all the experiences of humanity before the inevitable, whatever that may be, occurs.

In this way, the doubling highlights Deckard's own singularity. When with his wife, he feels "[C]onscious of his own aloneness" (p. 167). Deckard, not an android, cannot be duplicated or replaced, although Rachael and the other androids do their level best to make him doubt this. Deckard has sex with Rachael to prove to himself that he is human, that he is singular. However, Rachael, who is not singular, *can* be replaced, and the parts of her that thought she was human viciously fight against that replacement. She tries to assert her singularity, with Deckard, and rather fails:

> Gathering a giant white bath towel about her, Rachael said, 'Did you enjoy that?'
> 'Yes.'
> 'Would you ever go to bed with an android again?'
> 'If it was a girl. If she resembled you,' [p. 181].

By reminding Rachael of her non-uniqueness, Deckard hurts and angers her, and she hits back against that pain.

After they finish having sex, she comments, "You look so sad" (p. 182). While Deckard mourns her eventual demise, due to the nature of Nexus-6 androids, she thinks of her self-preservation and the preservation of her species:

> 'You're not going to be able to hunt androids any longer....'
> He stared at her.
> 'No bounty hunter ever has gone on,' Rachael said. 'After being with me....'
> 'I see,' Rick said. He felt numb. Completely. Throughout his entire body, [p. 182].

We, the reader, know that she is lying. From our first meeting Rachael Rosen, in the Rosen Association, we know that she, in this particular instantiation, was created to represent the association, that her memories of before are implanted, that she thought she was human until just a few hours ago, and there would have been no time in which she could have seduced any other bounty hunters vis-à-vis her android status, corrupting their ability to hunt. Deckard should know this too, but with the alcohol, sex, and exhaustion clouding his mind, he misses this fact.

However, Rachael's counterpart, Pris, knows of this doubling—having previously given her name as Rachael when first meeting Isidore, and attempts to use it to her advantage when Deckard shows up to retire the remaining Nexus-6 androids:

> And then he saw that it was not quite Rachael.
> 'For what we've meant to each other,' the android said as it approached him, its arms reaching as if to clutch at him. The clothes, he thought, are wrong. But the eyes, the same eyes. And there are more like this; there can be a legion of her, each with its own name, but all Rachael Rosen—Rachael, the prototype, used by the manufacturer to protect the others. He fired at her as, imploringly, she dashed toward him....They can follow me with Rachael Rosens until I die, he thought, or until the type becomes obsolete, whichever comes first, [p. 204].

Deckard does retire Pris, but soon after realizes that his ability to work is compromised, just as Rachael predicted. The novella ends with him in a quasi-catatonic state, a result of his obsession with Rachael conflicting with his reality: home, wife, job, and endless pursuit/conflict with his own humanity. This section has shown Deckard's motivations for sleeping with and falling in love with Rachael; however, it leads us to ask a question in regard to the technology invented by Dick in the novella: why are the androids gendered at all?

The Gender Binary, Traditional Gender Roles, and Early Robotic Technology

We can ask the same question about humans: Why are humans gendered at all? In order to answer this, we must make the distinction clear between gender and sex. Sex refers to a set of physical, biological properties that the scientific and medical communities have collectively decided are different enough from each other to form categories. The categories are, generally, referred to as female, male, and intersex. These categories are based on sets of traits that can be genetic (as in combinations of X and Y chromosomes), or physical (as in combinations of primary and secondary physical sex characteristics). They are still used because the good majority of human babies do fall within definably binary states—they have a penis or a clitoris, they have testicles or ovaries, they have XX or XY. The catch-all category, intersex, is used medically for those babies who do not—those who have a combination of traits, or traits that fall outside of what the medical and scientific communities call 'normal.'[4] At birth, the doctor performs a physical examination and makes a judgment call to assign sex, which is why the preferred terminology for people who are transitioning is not 'born male/born female,' but rather assigned male or assigned female at birth.

What, then, does this have to do with gender? Gender identity represents your internal feelings of masculinity or femininity or anything in between, and gender expression is how you manifest those feelings in your external presentation. Gender expression can directly influence how you are seen by other members of your immediate community—if you were assigned female at birth, and you identify with female gender identity, and you show outward expressions of femaleness and femininity, then you are likely to be understood on first glance as a girl or a woman (depending on your age). In this case, as your outside presentation is in agreement with your internal mental state, you are a cis-gendered female human. While much work has been done on the cultural reasons for the primary binary, male/female, this rather neatly ties back to Levi-Strauss and his work on myths and balanced oppositions—we, as humans, across time, like to put things in opposing pairs: raw and cooked, dark and light, night and day, male and female (1962/1966, 1955). Life just seems to make more sense to us if we can do a two-part compare and contrast.

However, as shown in the medical example, and in the discussion of individuals who are or have gone through the process of transitioning, the binary does not completely approximate reality. There are people who do not fit neatly into one of the two checkboxes. Robots would seem to be an obvious addition to this third category, this non-binary, intersex category. With no need to make them gendered at all, even if they are humanoid, we still gender robots based on what they do and how they look.

Jennifer Robertson gives this a thorough examination in her book *Robo Sapiens Japanicus: Robots, Gender, Family, and the Japanese Nation* (2018). In her words, ...robot gender effectively reproduces a sexist division of gendered labor among humans and humanoids alike" (p. 82). While humans have decided on biological categories of traits that determine sex (Agarwal & Wesp, 2017), and attempted to apply those categories to previous cultures (p. 100), the seemingly natural extension of those categories to robots betrays our own flaws and preoccupations with the roles people play in society. These roles become ingrained as early as childhood, and children ascribe sex and gender to their dolls and toys (Turkle, 2011). In an experiment with a toy called 'My Real Baby,' children recognized gender/sex, calling the baby her instead of it, and some were tender and caring, expressing a desire not to hurt her and protecting her from other children. Turkle reports, "[I]n the pandemonium of Scot and Alana's playgroup, My Real Baby is alive enough to torment and alive enough to protect" (p. 48).

And so we circle back around to the question of why gender robots at all? They are not human, they are metal and plastic, circuitry and wires, software and programming. They are, or should be, inherently genderless. The answer comes in the idea of those societal roles—what we do and who does what are ideas to which we have given gender. When a human creates a robot to fill a certain role, to perform certain tasks, the gender that has been assigned to those tasks often goes with that creation (Robertson, 2018, p. 82). That is not to say that these roles have no overlap, or remain tightly constrained and defined. All societies contain examples, whether singular and memorable, or widespread and celebrated, of people who gender-bend and/or reject the role society assigned them (Pêcheux, 1975). Sometimes people do this as an accepted and non-deliberate method of reflecting their authentic internal self, as in the two-spirit 'third' gender recognized by many Indigenous American groups, or the hijra of India; this is just who they are. Other times, it is a directly counter-cultural act, a way to set themselves apart in reaction to an oppressive or rigidly defined set of roles that they do not fit in, as in the case of many androgynous presenting people in the USA and UK during the 1970s and 1980s. Yet the roles remain, and we continue to use them when it suits us.

One way it suits us is in the category of sex robots. Here we have another dichotomy: robots meant to be used for sex, and robots meant to be used for other tasks (which may or may not be gendered). In the novella, the androids are gendered based on client preference (Dick, 1968/2017), and, in reality, that can be the case as well: "The vast majority of robots are also gendered, sometimes ambiguously in the case of robots whose purchases have the opportunity to name, and thus 'gender,' the robot" (Robertson, 2018, p. 82). However, they did not start out deliberately so.

Cog, a robot designed and built by MIT scientist Rodney Brooks in the 1980s, does not have a face. It has brackets, servo-motors, wires, camera lenses, and articulating machinery, but it does not have a face. However, in a psychological process called 'pareidolia,' in which the human mind sees patterns whether or not those patterns are physically present, we ascribe a face to Cog—it has bilateral symmetry, which makes this process easier, but we would do so regardless. On his website, when asked about if Cog is a boy or a girl, Brooks states, "No. The robot has no gender, and we try to use the pronoun 'it' whenever it comes up. We also carefully construct our sentences to avoid third person references" (Brooks, n.d., FAQ). No matter how humanoid, however, Cog is not a sex robot, and no one would mistake it for one—its non-genderedness is not off-putting when anthropomorphizing it, but it would be in terms of performing sex acts.

For example, when we look at sex dolls—even blow-up dolls, which are incredibly primitive—they are distinctly gendered. The males and females have different proportions, the waists are different sizes, there are breasts on the female dolls. The more recent, and more distinctly human sex dolls, such as those sold by RealDoll, Sinthetics, and other manufacturers, are either deliberately binary, or purposefully can be made to be intersex, but the largest demand is for a doll that reifies the binary—that checks male or female to the human eye. Hidetoshi Akasawa of the Science University of Tokyo has created a robot head with the ability to create and eventually react to human facial expressions. Shown in the book *Robo Sapiens: Evolution of a New Species* (2000) (no relation to the aforementioned book by Robertson), the head is decidedly female, and although she is described as "flashing a smile" (p. 73), the expression reads more like an grimace to me. The expressions for RealDoll and those made by Sinthetics are more constrained than the robotic technology demonstrated by Akasawa's nameless bot, constrained as they are by the necessity for soft, flexible, tooth-free oral cavities.

This brings us back to the idea of the roles being performed. With sex dolls and sex-bots, those roles are generally to be a specific type of companion, to have a specific type of sexual compatibility. Male sex dolls are equipped with the same flexible oral and anal cavities, and with a penis attachment. Female dolls are equipped with oral, anal, and vaginal cavities, and customizable labia (Sinthetics; RealDoll). They are not required to smile, emote, or mimic human behavior, because they are made for sex. The people who purchase, own, collect, or use the dolls, however, do attribute personalities to them, much as the androids in *Do Androids Dream of Electric Sheep?* (Dick, 1968/2017) attribute personalities to themselves, when it suits them to do so. So, if we give robots humanity, and therefore gender, how do cyborg or cybernetic additions to our own bodies, or our use of sex dolls or sex robots, confound our ideas about ourselves?

What Is the Opposite of a Manifesto? Daily Cyborgs, Daily Lives

In 1991, when Donna Haraway wrote *A Cyborg Manifesto*, it was revolutionary—a new paradigm emerging, we, all of us, though women specifically, were cyborgs—a merging of person and other, where other was usually machine. This has not changed in the last 27 years, other than to be more firmly reinforced in our consciousness. We use mechanical interventions in multitudinous aspects of our daily lives, from relying on Siri™ or Alexa™ to organize our schedules and find us directions, to the integration of the Roomba™ into our households, to new innovations in various types of health care. This also applies to our relationships with sex dolls and sex robotics. While outside of the norm, such things are no longer taboo, and are on the verge of becoming mainstream.

However, when Dick was writing in 1968, such things were not even dreamt of by the manufacturers of today. As of the time of this writing, it has been 51 years since the publication of *Do Androids Dream of Electric Sheep?*; 38 years since IBM launched the first personal computer; 20 years since the first Blackberry device hit the market; and 12 since the launch of the first iPhone. The progression of this technology is mathematical, following the predictable technology development curve. Ideas formed in the minds of sci-fi writers drive innovation in technology, and technological progress grows at a predictable, though far from steady, rate.

This explains the sparse literature on the topics of robotics and cybernetics during the era in which Dick wrote; he was literally ahead of the curve. We see more scientific studies beginning in the 1980s, when computerized medical technology catches up with science fiction. One such study was conducted by Goldberg, Weller, and Blittner, published in 1982, and concerns patient interactive biofeedback during EMG testing. EMG, or electromyography, uses electricity to stimulate muscular nerves, and is used to diagnose musculo-nervous or neurological conditions such as multiple sclerosis and peripheral neuropathy. The EMG test the authors used was called the "Cyborg Distributive Processing System," in reference to the fact that the combination of computer technology and the human participant created a present, if temporary, cybernetic organism.

Yet are cyborgs 'human enough'? This question only occurs after Haraway's essay. Before that watershed year of 1991, the conception of cyborgs, cybernetics, or non-flesh-material replacements for human parts remained foreign to our conception of humanity, despite pacemakers and iron lungs being ubiquitous as early as the 1950s. Rather than being contradictory, however, this meshes well with cultural attitudes toward such assistive devices, which necessitated physical and social isolation. Those who used iron lungs were immobilized, and only had interaction with those people who came close voluntarily. The first pacemaker was "the size of a small cathode ray tube (CRT) television, and the patient had to push it around like a shopping cart" (Siemens, 2019). Instead of integrating into the lives and physicality of the patient, such devices overwhelmed the human body, and created distance between the person and their fully flesh counterparts.

Post-Haraway's essay, however, a certain theoretical turn embraced cyborgs and cybernetics as a useful tool for examining our relationship with medicine and health, as well as sex. One very clear example of this, a book called *Cyborg Babies: From Techno-Sex to Techno-Tots* by Davis-Floyd and Dumit (1998), as anthropologists who consistently critique the American system of childbirth, incorporates narratives from technologically aided conception to the 'non-natural' permutations of childbirth in female centered essays. Such essays show how, gradually, over time, cybernetic technology becomes integrated into our daily lives. Yet, we do not live in a post-tech world—assistive devices still 'other' people. There remains a stigma attached to assisted conception, or to using a wheelchair, or to having sex with a non-human humanoid. In fiction, this otherness occasionally manifests as racialization of characters.

Rachael Passes—Judaism as a Metaphor

Orientalism, by Said (1978), can be seen as an indictment of paternalistic, colonial ideas about the so-called Western versus the so-called Eastern mentalities, particularly those mentalities that involve the saving, civilizing, or taming of the East by the West. Such mentalities make an easy literary device, however—a way to quickly and clearly set up a character as different, or other. While Said was mainly discussing the Far East (parts of Asia including India, Japan, and China), the conceptualization of East/West or us/them does not conform to those limitations. In *Do Androids Dream of Electric Sheep?* (1968/2017), the Rosen Association, and Rachael Rosen in particular, are set apart by virtue of their distinctly Jewish names.

Rosen is a German-derived Ashkenazi Jewish name, and Rachael is a biblical Jewish/Israelite name meaning 'ewe,' a female sheep. By comparison, Rick and Deckar(d) are both of Dutch roots, which sets up an implicit East versus West dichotomy in the mind of the reader, though in this case it is the Middle East, not the Far East of Said's analysis. This taps into two competing larger narratives of the time—the American hatred of Nazism and quasi-welcoming of Jewish contributions to the American way of life, and the historical outsider status of Jews which has recognized beginnings in the segregation of Jews during the Renaissance, Middle Ages, and Early Modern times. While rooted in stereotypes, the characterization of the Rosen Association reinforces ideas about Judaism which were deeply held in American thought in the 1960s, specifically that Jews were power-hungry, somewhat secretive, good at finances, and looking to infiltrate society (Brodkin, 1998).[5] Rachael epitomizes this.

In discussing the Uncanny Valley, as seen in Chap. 1 of this book, there is a drop-off after which something is too human, approximately 85% (Chu, 2015, p. 76). In order to determine what traits are human and which are non-human or too human, we stereotype (p. 80). A human should look or act a certain way, and outside of that way, we declare them non-human or too human. And in the case of Rachael, the question of her humanism and the question of her Western-ness are one and the same. The answer to the question lies in her lack of overt emotion, and her symbolic nature, both of which incite desire in Deckard.

There are three types of symbolism embodied by Rachael: the racial other who almost passes, the exotic object of desire, and the non-human who almost passes. Whereas Deckard initially desires a real sheep, to replace his electric one, that desire becomes transferred to Rachael, the non-Western, non-human, exotic, who can almost pass as human and as Western. Rachael becomes the ewe of her name, she becomes that which is desired, and her last act in the novella is to kill that which she would replace—not Deckard's human wife, Iran, but the physical, real, live goat he purchased to maintain his status. With cybernetics and robotics in their infancy as fields of creation and study, physical robots embody these symbols as well. The next chapter will show how these themes develop, as we move from the novella to *Blade Runner* (Scott, 1982), the movie.

Notes

1. This technology is not so far out of the realm of possibility… (Spiegel & Lambert, 2019).
2. I have chosen not to address the rampant ableism in the novella. However, it does bear noting that several terms used in regard to chickenheads are offensive by 2018 standards, and were not necessarily seen as such in 1968. My description of chickenheads and subsequent reference to the traits as expressed in the novella does not indicate approval of such terminology or prejudices.
3. Or, as stated in Monty Python and the Holy Grail: "Come and see the violence inherent in the system! Help, help, I'm being repressed!" (Gilliam & Jones, 1975).
4. 'Normal' has a medical definition and a cultural definition, and here I am using only the medical definition. There is nothing abnormal about being intersex, though it is statistically less likely than physically reifying the binary.
5. Again, this does not express my own thoughts, and is merely an explanation of the stereotype.

Bibliography

Agarwal, S., & Wesp, J. (2017). *Exploring sex and gender in bioarchaeology*. Albuquerque, NM: University of New Mexico Press.

Bordo, S. (1987). *The flight to objectivity: Essays on Cartesianism and culture*. Albany, NY: State University of New York.

Brodkin, K. (1998). *How Jews became white folks and what that says about race in America*. New Brunswick, NJ: Rutgers University Press.

Brooks, R. (n.d.). *Humanoid Robotics Group FAQ*. Retrieved from http://www.ai.mit.edu/projects/humanoid-robotics-group/cog/faq.html

Chu, S.-Y. (2015). I, stereotype: Detained in the uncanny valley. In D. S. Roh, B. Huang, & G. A. Niu (Eds.), *Techno orientalism: Imagining Asia in speculative fiction, history, and media* (pp. 76–88). New Brunswick, NJ: Rutgers University Press.

Davis-Floyd, R., & Dumit, J. (1998). *Cyborg babies: From techno-sex to techno-tots.* New York, NY: Routledge.

Dick, P. K. (1968/2017). *Do androids dream of electric sheep?* New York, NY: Del Ray.

Foucault, M. (1984). Excerpts from discipline and punish. In P. Rabinow (Ed.), *The Foucault reader.* New York, NY: Pantheon Books.

Gilliam, T., & Jones, T. (Producers), & Forstater, M., & White, M., (Directors). (1975). *Monty Python & the Holy Grail* [Motion picture]. UK: Michael White Productions.

Goldberg, J., Weller, L., & Blittner, M. (1982). Cognitive self-control factors in EMG biofeedback. *Biofeedback and Self-Regulation, 7*(4), 545–551.

Levi-Strauss, C. (1955). The structural study of myth. *The Journal of American Folklore, 68*(270), 428–444.

Levi-Strauss, C. (1962/1966). *The savage mind.* London, UK: The University of Chicago Press.

Menzel, P., & D'Aluisio, F. (2000). *Robo sapiens: Evolution of a new species.* Cambridge, MA: Massachusetts Institute of Technology.

Pêcheux, M. (1975). The subject form of discourse in the subjective appropriation of scientific knowledges and political practice. In H. Nagpal (Trans.), *Language, Semantics, and Ideology* (pp. 155–170). New York, NY: St. Martin's Press.

Pryor, C., Perfors, A., & Howe, P. D. L. (2018). Even arbitrary norms influence moral decision making. *Nature Human Behavior, 3*, 57–62.

RealDoll. (n.d.). *Home page, testimonials, and product information.* Retrieved from http://realdoll.com

Robertson, J. (2018). *Robo sapiens Japanicus: Robots, gender, family, and the Japanese nation.* Oakland, CA: University of California Press.

Said, E. W. (1978). *Orientalism.* New York, NY: Pantheon Books.

Scott, R. (Director), & Deeley, M. (Producer). (1982). *Blade Runner* [Motion Picture]. USA: The Ladd Company.

Siemens. (2019). *Siemens history.* Retrieved from https://www.medmuseum.siemens.com/en/stories-from-the-museum/herzschrittmacher

Sinthetics. (2012–2019). *Testimonials and product descriptions.* Retrieved from http://sinthetics.com/

Spiegel, A., & Lambert, J. (2019). *Are we ready for an implant that can change our moods?* Retrieved from https://www.npr.org/sections/health-shots/2019/03/29/707883163/are-we-ready-for-an-implant-that-can-change-our-moods?utm_source=facebook.com&utm_medium=social&utm_campaign=npr&utm_term=nprnews&utm_content=20190329&fbclid=IwAR1PAIfxffj23w7RzA0BgduJTtpL5HtHT15Zg-LTOxemEdjlKomJT3Juz10

Turkle, S. (2011). *Alone together: Why we expect more from technology and less from each other.* New York, NY: Basic Books.

CHAPTER 3

Incep Dates and Pleasure Models: Death, Life, and Love in *Blade Runner*

Abstract This chapter examines the first *Blade Runner* movie, Ridley Scott's adaptation of the original novella (1982). In *Blade Runner*, Replicants become undetectable, undifferentiated visually, from humans. The newer models can pass the Voight-Kampff test—a test which distinguishes true emotional responses from implanted ones. This perfect mimicry is our next iteration of how we desire sex robots to behave, and it is desired well in advance of the possibility of building such bots. It is in this iteration of the story where two theoretical questions appear—that of the racialization of the characters, which is far more apparent in a visual media, and the humanity of death—that how being able to die, and the fear of it, humanize the movie's Replicants.

Keywords *Blade Runner* • Death • Race • Gender roles • Sex • Replicants

Synopsis

Blade Runner (Scott, 1982) is the movie adaptation of Dick's novella *Do Androids Dream of Electric Sheep* (1968/2017). Set in post-apocalyptic Los Angeles in November 2019, it covers Deckard as he hunts down four of six escaped Replicants.

R. Gibson, *Desire in the Age of Robots and AI*,
https://doi.org/10.1007/978-3-030-24017-2_3

The movie begins, showing one Replicant, Leon Kowalski, played by Brion James. A Blade Runner administers the Voight-Kampff test to the Replicant. Leon has infiltrated the Tyrell Corporation (the movie's version of the Rosen Association) and works there undercover as a waste disposal engineer. A few questions into the test, Leon pulls out a gun and shoots the Blade Runner. Deckard, played by Harrison Ford, is brought in as a replacement.

We meet Deckard on the streets of Los Angeles. The city has been transformed into an Asianized landscape, with one building displaying a light up billboard with a Geisha, the neon signs in various Asian languages, and the vendors and citizens mostly Asian. Deckard mentions that the language is "city-speak, guttertalk, a mishmash of Japanese, Spanish, German, what have you" (Scott, 1982). We get a glimpse of Deckard's previous life—he is retired from being a Blade Runner, and his ex-wife found him to be "sushi…cold fish" (ibid.). Gaff, a cop, brings Deckard to the LAPD building to meet with Lieutenant Bryant, a coarse, crass, stereotype of a cop who calls Replicants 'skinjobs,' a slur akin to the n-word.[1] The LAPD building has the shape of a panopticon, circular and arranged around a central column. Bryant tells Deckard that there are four escaped Replicants who want revenge on Tyrell—Leon, Zhora, Pris (played by Daryl Hannah), and Roy Batty (played by Rutger Hauer). During their escape, they and two others managed to kill 23 humans on their way to get to Tyrell. Each Replicant has a type—Batty was built for combat, Leon is not very smart but is quite strong, Zhora was built to seduce and then murder, and Pris is a "basic pleasure model." Their 'incep dates' or the date they were made, range from 2016 to 2017.

The incep dates signify something else: a death date. The movie takes place in November 2019, and Tyrell built a 4-year life span into the Nexus-6 Replicants to avoid their inevitable emotional growth—Replicants with emotions are dangerous, they care about their own lives and about their own deaths, and they cannot be controlled, so they kill, as Leon, Zhora, Pris, and Roy are doing. Bryant orders Deckard back into service, and Deckard tries to refuse. Bryant states, "if you're not a cop, you're the little people," as Gaff folds an origami chicken and sets it down beside Deckard.[2] Realizing he has little choice in the matter if he wishes to keep his dignity or masculinity intact, Deckard goes to the Tyrell building to test out the Voight-Kampff on the Nexus-6 model.

He meets Rachael, played by Sean Young, for the first time. The owl is there, and Deckard asks if it's fake—she admits that it is, and that it was expensive. Eldon Tyrell enters and requests a demonstration of the Voight-

Kampff test on a human first, to calibrate it. He volunteers Rachael. The details of the test are the same as in the novella—Deckard asks her about various scenarios, and she responds almost verbatim, including the line, "Is this testing whether I'm a Replicant [android], or a lesbian, Mr. Deckard?" (ibid.). However, in this situation, he uses 'over a hundred' questions when it usually takes far fewer. Tyrell eventually asks her to step out, and Deckard asks:

> "She doesn't know?"
> "She's beginning to suspect, I think."
> "Suspect?! How can it not know what it is?"
> "…More human than human is our motto. Rachael is an experiment, nothing more," (ibid.).

In the meantime, Leon has tipped off Roy that their cover has been blown, and they follow a three-pronged strategy for escaping Deckard while still finding their way to Tyrell. Zhora maintains her cover by dancing at a local club, while Leon and Roy try to track down people connected to the Tyrell Corporation so they can find out if the shutdown dates are absolute. After being told to ask J.F. Sebastian, Pris goes to befriend him. Sebastian is the movie's version of J.R. Isidore, but rather than being a chickenhead, he has a genetic defect—Methuselah Syndrome—that makes him age very quickly.

Rachael traces Deckard to his apartment. They argue about whether or not she is a Replicant. She shows him photos of herself and her mother, outside their house. Deckard counters that those photos, those memories, are false. They do not belong to her; they belong to 'Tyrell's niece.' She weeps with the realization that he is right. He desperately tries to walk back what he said, to comfort her, but he fails, and she runs off. Not only has she developed emotions, he has developed feelings for her.

Pris has moved in with Sebastian. In addition to being a genetic developer for the Tyrell Corporation, Sebastian makes toys and calls them his friends. Pris invites Batty to the deserted house in which Sebastian and his toys live. Meanwhile, Deckard has tracked Zhora to a club called Taffy Lewis's. Taffy's is a strip/burlesque club, and Deckard invites Rachel to join him for a drink. Looking shocked, Rachael replies, "I don't think so, Mr. Deckard. That's not my kind of place" (ibid.). Zhora goes on stage as 'Salome' dancing with a snake—highly symbolic, as the announcer states "watch her take the pleasures from the serpent that once corrupted man!" (ibid.). Deckard apprehends Zhora in her dressing room, then chases her when she becomes aggressive toward him, eventually killing her in the street. Bryant and Gaff show up to commend him on the execution of his duties, reminding him that there are four Replicants left to retire. He replies vehemently that there are three—Leon, Pris, and Roy Batty. No, says Bryant, there's a fourth. Rachael has disappeared from the Tyrell Corporation, and is now considered illegally at large. She is standing across the street at the time, and she hears this. Leon shows up and goes for Deckard, so Rachael shoots him with Deckard's gun, which he had dropped during the fight.

They go back to his apartment, where she takes down her hair, letting the soft waves flow around her face, and plays the piano—she wants to see if she can or if the ability is only an implanted memory. He kisses her. She goes to leave, but he blocks her way. He throws her against the wall and kisses her again:

> Say 'kiss me.'
> Kiss me.
> 'I want you.'
> I want you.
> Again.
> I want you. Put your hands on me, (ibid.).

The scene switches to the Bradbury Building, where Pris wakes Sebastian and introduces him to Batty. Pris and Batty aggressively kiss, confusing Sebastian (and the viewer—after all, they are technically 'siblings'). After Batty threatens him, Sebastian agrees to take him to see Tyrell to discuss their shutdown dates.

At the Tyrell Corporation ziggurat, Sebastian watches as Batty asks if anything can be done to expand their expiration dates. He and Tyrell go back and forth over methods, but in the end there is nothing that can fix the Replicants—they are going to die. And so, Batty kisses Tyrell on the lips and kills him by pressing his thumbs into his eyes. Sebastian runs off in shock, and Batty follows, eventually killing him as well. He then returns to the Bradbury for Pris.

Deckard has tracked Pris and Batty to the Bradbury, and after a brief struggle, he retires Pris. Batty then arrives and chases Deckard through the building, taunting him, hurting him, circling back around to mourn Pris and kiss her one last time. In a weirdly and unexplainedly sexual turn of phrase, Batty says to Deckard (about his gun), "you'd better get it up, or I'm gonna have to kill you!" (ibid.) even though killing Deckard has seemingly been his intention all along. Batty's capabilities are also well up to the task—he remains obviously physically superior throughout the entire chase.

Deckard drops his gun while climbing part of the building, and goes to the roof with the intent of jumping to the next building to escape. He slips and dangles from the building. Batty makes the jump easily, and as Deckard starts to fall, he catches him and hauls him back up to the roof, saving his life. Batty is actively dying at this point, and he knows it—perhaps all the damage he has recently taken has accelerated his end date:

> I've seen things you people wouldn't believe. Attack ships on fire off the shoulder of Orion. I watched C-beams glitter in the darkness at Tannhäuser Gate. All those moments will be lost in time like tears in rain. Time to die, [ibid.].

Gaff shows up and tells Deckard that Rachael will not live, "but then again, who does?" (ibid.). Deckard goes back to his apartment, terrified that he will find her dead, that Gaff has killed her, but she lives, asleep in his bed. Deckard kisses Rachael awake, and says:

> Do you love me?
> I love you.
> Do you trust me?
> I trust you, [ibid.].

They flee the building and she kicks over an origami unicorn; Gaff has indeed been there, but chose to leave her alive. As they fly off into the sunrise in Deckard's car, the voiceover tells us that Tyrell designed Rachael to be special, and she does not have an expiration date.

The Undetectable Replicant: Why Must We Mimic Humanity?

Although Wosk (2015) gives us many reasons why we strive for perfection in the robot woman (see Wosk, Chapter 5), humanity is often defined by our imperfections (Levy, 2007, pp. 144–150). So, to be undetectable, the Replicants must walk a tightrope between human-because-flawed, and obviously-android-because-flawed-in-the-wrong-way. Much like real robots, all of the Replicants in *Blade Runner* (Scott, 1982) fail to sustain their disguises.

They cannot pass the Voight-Kampff test, and they do not comprehend humans enough to fit in during speaking with or interacting with humans. Furthermore, the medium of film gives rise to a distinct trait, not mentioned in the novella. Under certain lighting conditions, the Replicants eyes glow like the sun hitting a mirror. They are opaque and reflective, rather than being transparent and showing the void of the pupil, as human eyes do.

All of the Replicants in *Blade Runner*, apart from Rachael, show their otherness to varying degrees, and some embrace their failure to blend-in in rather spectacular ways. This relates to the various 'types' of Replicants that each one personifies. Leon—the engineer, the cleaner; Roy Batty (the only Replicant other than Rachael who is given a last name), the leader; Zhora, a pleasure model who is also combat trained; and Pris, the basic pleasure model. Leon and Zhora give themselves away almost as soon as they are challenged—Leon shooting the Blade Runner giving him the Voight-Kampff test, and Zhora attacking Deckard when he begins to question her in her dressing room. However, the movie feels as though Leon and Zhora are present to propel the plot more than having a substantive role. Often they serve to bring Deckard to a place where he needs to be, and their eventual demise, within minutes of screen time of each other, reunites Deckard with Rachael after she has fled his apartment and turned down his illicit proposition. Their purpose and their dialogue could have been divided up between other characters, or indeed, left out entirely, and the movie would have proceeded just fine. The important Replicants are Roy Batty, Pris, and Rachael.

Roy and Pris mimic humans perfectly when they choose to, though Pris chooses to do so more often than Roy. This, again, goes back to their types. Roy is the leader. He can effectively lead the Replicants and interface with humans, because he was built to be more intelligent and to be more in-tune with human norms. You must understand rules before you can break them, and Roy understands them perfectly. He breaks them to great effect as well. One such instance involves a trip to the 'Eye World' where a technician creates Replicant eyes. It takes a while for the technician to realize that Roy is a Replicant, before recognizing his own handiwork in Roy's eyes. Throughout the questioning, Roy uses charm, persuasion, and implicit and explicit threats to get information from Chew, the Tyrell Corporation eye technician, who tells him that he must speak to Tyrell himself in order to find answers. Chew says that J.F. Sebastian can take Roy to Tyrell.

The way to Sebastian, however, is Pris. Pris, being a pleasure model, was built to know what people (implied: heterosexual men) want. She tracks Sebastian to his home and hides herself in a pile of trash, pretending to be scared and startled into flight when he arrives home. Claiming to be lost, homeless, and alone, she seems gentle and harmless. Sebastian lets her inside. She becomes a 'human' connection to him, as he states that the only friends he has are those he makes—the robotic toys that fill his apartment. Her friends, she says, are in town, but they are not worried about her; she will let them know where she is the next day.

These examples tie back to what Levy (2007) stated about the ten factors of attraction. Pris definitely flirts with Sebastian, leading him to experience several factors (similarity, desirable characteristics, reciprocal liking, filling needs, arousal/unusualness, specific cues, isolation, and mystery). Sebastian isolates himself in a part of town which houses few others, in a huge building that he alone lives in. He becomes intrigued by Pris' similarities, as she is lonely, singular, and seemingly gentle and harmless. She creates reciprocal liking, by linking them together right away with the dialogue line "we scared each other pretty good, didn't we?" (Scott, 1982). However, although Sebastian's toys are the automata of history (Voskuhl, 2013), there are two that are semi-sentient, which do not trust Pris (they are played by Kevin Thompson [as Bear] and John Edward Allen [as Kaiser], who were both actors with dwarfism). In this we can see how uncanny being too similar to something becomes; their distrust stems from a more visceral knowledge of Pris' nature than Sebastian has access to.

Yet, soon Sebastian does know. This occurs when Pris and Roy reunite the next day. Sebastian wakes to Pris examining his 'friends,' though we, the audience, see that she has sniffed him. She has airbrushed her eyes, and there is now a black bar across her face, suppressing some of the emotional affect-mimic of which she is capable. Through it, her eyes flash Replicant. She realizes that something is wrong with Sebastian, and he tells her he has Methuselah Syndrome, a genetic condition that causes rapid aging and keeps him from moving off-world. Despite this, she tells him, "I like you, just how you are" (Scott, 1982). Roy steps out of the shadows, and Pris greets him. They aggressively kiss. This seems to contradict her courtship of Sebastian; however, she remains true to her type—the first person she must please is her leader, so she turns her attentions to Roy as soon as he wants them.

Rachael, it turns out, is the anomalous model. With implanted beliefs that she is human, with real memories (though not her own memories, they *are* real), and with no pre-programmed shutdown date as Deckard eventually finds out, Rachael can almost pass for human. Replicants are not supposed to feel, but she does. They are not supposed to care, but she does. And of the two of them, Deckard is the more violent, the less predictable.

Gender Roles Again?: Deckard's Lack of Emotion, Rachael's Humanity, and the Question of Replicant Gender

However, Deckard's volatility, Rachael's passivity, taken at face value, could merely be a resurgence of the gender binary we saw in the novella. This deserves to be unpacked, because while on the surface we can reduce their roles to a series of tropes, Scott does appear to be doing something unique with the characters rather than simply having them reflect the gender mores of the time. Let us take a closer look at the behavior of each.

When we first meet Deckard, we see that he has a past: "Sushi…that's what my ex-wife called me. Cold fish" (Scott, 1982). This line is part of a voiceover narration given by Harrison Ford in the original theatrical release of the movie. Ford has spoken out about his reluctance to do the voiceover, and that reluctance very obviously caries through in how it is delivered, so while I might otherwise be analyzing vocal tone (as I will do for characters from *Blade Runner 2049* (Villeneuve, 2017)), I will not be doing so here. He is bored, and it shows. However, this line gives us a clue into Deckard's overall personality—he is stolid, unemotional, uncaring, and does his job. The bored tone suits the neo-noir feeling of the movie.

He is also, at the beginning of the movie, retired. This word contains a double meaning: in one sense it can be taken quite literally—his job was to be a Blade Runner, and the LAPD no longer employs him as one, so he has retired. In another sense, however, 'retired' is the word used to refer to having killed a Replicant, both in the novella and in the movies. Metaphorically, this can also refer to Deckard. He is older, divorced, no longer productive, and has a quite pronounced alcohol problem. He may as well be 'retired.' Both Bryant and Gaff remind him of this, with Bryant saying the line, "If you're not a cop, you're little people" (Scott, 1982), and Gaff folding the origami chicken. Deckard, easily goaded with memories of his former importance and knowing that he truly is necessary to the mission of 'retiring' the escaped Replicants, agrees to reenter service.

Throughout the movie, Deckard shows stereotypically masculine personality traits. He becomes sexually aggressive with Rachael; he is, in Bryant's words "a goddamn one-man slaughterhouse" (ibid.) when it comes to killing Replicants, and he continues to struggle with the idea that he has feelings at all, let alone for Rachael, the final Replicant he is supposed to kill. Furthermore, he sees Replicants as very masculine—he thinks they are supposed to be aggressive, kill easily and without remorse, and have no real feelings. Rachael, who does have feelings and subtly shows them, confounds his expectations, though the rest of the Replicants do not.

As discussed in the novella, the primary goal of the Replicants must be self-preservation. This continues in the movie, as they (or at least Roy) are aware of their own shutdown dates. And also as before, their plans almost immediately fall apart—they have very little ability to plan for the long term, to play for the endgame rather than merely moving the game piece immediately in front of them. Each Replicant, with the exception of Rachael, when given the opportunity to continue the deception or attack, chooses to attack. They also do not fulfill completely the stereotypical gender roles. While it is true that Zhora and Pris have pleasure-model components, with Pris being the total pleasure model and Zhora being a combination of pleasure-bot and combat model, both are cavalier about violence, aggressively sexual when they need to be, and they use all aspects of themselves to succeed in their mission to maintain a living. They exemplify 'Replicant' rather than 'female.' The difference here lies in Rachael's lack of knowledge, and in her eventual acceptance that her entire worldview must change with the understanding of who and what she is.

Rachael, however, does exemplify 'female,' particularly 'human female.' This ties into the question of what we desire, and why we would even give robots or Replicants gender at all. According to Campbell (2010), the ability to see ourselves in the formation of cyborgs and cybernetic technology represents an ontological turn in human thinking and conception—we created a new way of being when we thought of our own bodies as something malleable enough to contain robotics technology. Accepting that we could be them (robots/cyborgs), we can also accept that they can be us (human). The Replicants are given gender because we do not yet desire a genderless world, and we have only begun to see a world where the border between robotics and living human flesh blurs.

Yet Rachael is desirable not for her Replicant nature, but for her human one. She is vulnerable, emotional, chaste, and delicate, while her fellow female-type Replicants are not. She does not overtly demonstrate the intense physical strength that the other Replicants have, though certainly it exists. Instead, she displays reciprocal desire. When prompted by Deckard, she tells him she wants him, she loves him, she trusts him, she wants his hands on her, she wants to kiss him (Scott, 1982). Her (relative) warmth thaws Deckard from the 'cold fish' we saw at the beginning of the movie, eventually leading to the pair of them spending approximately two years together after the conclusion of this movie's timeline. He does not just desire her for sex, but for companionship. This, then, leads to a further question: in the context of the movie, can we consider Rachael to be a sex-bot?

For His Pleasure: What Meets the Definition of Sex Robotics?

Danaher and McArthur (2017) and Levy (2007) are all pretty clear about the definition of sex robots: a sex robot is a created mechanical being that is used for the purpose of sex. However, in Rachael's case, it might behoove us to tease apart the various parts of that definition. There are two lines of inquiry here—is Rachael a created mechanical being? And is she being used for sex?

The opening crawl of the movie does state that Nexus Phase Replicants are robots (Scott, 1982). It calls Nexus-6 Replicants "superior in strength and agility, and at least equal in intelligence, to the genetic engineers who created them," (ibid.). Here, we have several of the requirements for robot-ness. They were created; they are non-human and have non-human abilities; they were even called robots. However, they are not completely mechanical; they are biomechanical. They breathe, eat and drink, and sleep.[3] This does not necessarily disqualify Rachael, or Replicants, from being robotic. Such actions—respiration, intake of solid and liquid supplemental material, and resting periods of low software activity—were well within the robotic capabilities of the 1980s (Menzel & D'Aluisio, 2000).

Therefore, we must move on to the next question—was Rachael being used for sex? Here, it is necessary to anticipate the sequel and bring in character continuity. In *Blade Runner 2049* (Villeneuve, 2017), the newest creator of Replicants Niander Wallace mentions that Replicant fertility is "Tyrell's final trick: procreation. Perfected, then lost" (ibid.). He later states that Deckard may have been bred or selected to mate with Rachael and produce a child. This refers back to the split between Replicant and human so evident in the novella—those who can breed, in the book, are human; androids cannot breed. Replicants cannot breed. Yet, this anticipates the answer to our question of whether or not Rachael was being used for sex.

In *Blade Runner* (Scott, 1982), Tyrell states that Rachael is "an experiment, nothing more." Was part of that experimentation a foray into the realm of fertility? Whether or not it was, Rachael's pregnancy must have come as a great shock to her and to Deckard. As far as either of them knows, she is as infertile as the rest of the Nexus-6 Replicants (though the movie and the sequel both imply that Rachael is not a Nexus-6, but the model after that—the only known Nexus-7 type). Whether or not her fertility was intended by Tyrell to be a feature rather than a bug, neither Deckard nor Tyrell mentions it during the movie. Deckard does mention other special features about Rachael—she was given very specific memories that may have been personally selected by Tyrell, and she does not have a built-in end-date. Therefore, we must conclude that her primary purpose as a demonstration model of the Nexus-7 line was what was completely evident: professionalism, compassion, the ability to almost pass for human, and, possibly, pleasure.

We can use this knowledge to address her sex-bot-ness. The Internet Movie Database (IMDB) lists 89 movies and TV show episodes that contain 'artificial females.' Considering the vast number of movies and TV episodes in total, that seems an extraordinarily low number; however, the list shows that even in most of the ones which are specifically made to be sex-bots (the list was not divided by sex-bot or just incidentally female bot), some aspect of companionship prevailed (imdb.com, 2016 [updated recently]). In the words of Danaher and McArthur (2017), "[S]ex robots have more going on [than sex dolls]." Deckard certainly wants sex from Rachael, and shows it with his (aggressive, masculine) actions—asking her to Taffy Lewis' bar (which she rightly interprets as a come-on for sex), blocking her way as she tries to exit his apartment, and demanding she ask him to kiss her and tell him she wants him.

Yet, he also wants companionship. Highlighting a deeply insecure and hesitantly tender personality, he wants her love and trust, too (Scott, 1982), though his way of asking is just as demandingly aggressive. She, a Replicant with true emotions, can provide what he is asking for, despite the fact that he knows that he will almost certainly outlive her. In this way, Rachael fits all of Levy's (2007) criteria for robotic companionship. We can conclude that she is a sex-bot, but not *merely* a sex-bot. Her desire to be human and the fact that she embraces her human side (unlike her instantiation in *Do Androids Dream of Electric Sheep?* (Dick, 1968/2017), who became violent and leaned into her android nature) and leaves with Deckard out of (expressed) desire, rather than only for her survival, make her characterization more complex than many of the other examples one can draw from IMDB's list.

From that list, which incidentally does not discriminate between robots and created humanoid beings, we see such one-dimensional characterizations as the *Stepford Wives* (both the original (Forbes, 1975) and the remake (Oz, 2004) based on the book of the same name (Levin, 1972)); 'Stella Mudd' a replica of a character's hated and nagging wife from the original series of *Star Trek* (Roddenberry, 1967); and Gigolo Jane from *A.I. Artificial Intelligence* (Spielberg, 2001). While these characters are strikingly built for one purpose, Rachael is multi-functional, and can express her own desires through subtle shifts in tone and expression that show her emotions. She weeps, and her voice shows natural inflections that indicate her sadness, disappointment, or stern refusal to be "that kind of girl" when Deckard asks her to Taffy Lewis' bar (Scott, 1982). She, it seems, has real human emotions, unlike the false and easily abandoned emotions shown by Pris.

The Rachael/Pris Dichotomy: Cartesian Dualism and Replicants

In *Blade Runner* (Scott, 1982), female Replicant Pris becomes a foil for Rachael. They are each other's opposites. With Pris representing the Nexus-6 pleasure model, and Rachael being the singular Nexus-7 realistic model, many of their similarities and differences need to be unpacked. While in the novella, Pris briefly claims to be Rachael and indeed represents a copy of the Rachael prototype, here their doubling is more metaphorical than actual—the actresses are not at all similar in feature or build, and disbelief is harder to suspend when one sees the dissimilarities on the screen.

This difficulty has been overcome before, in movies before this one was released (such as *The Parent Trap* (Swift, 1961)), using film tricks and mirrors to overcome difficulties of the actress being on screen as two characters at the same time. Scott could, indeed, have had Rachael and Pris both be played by Sean Young, thus more clearly sticking with the reality of the novella. However, the characters of Rachael and Pris in Scott's movie do not intersect—they do not share screen time at all. Thus, the decision to have two different actresses must have been one of characterization rather than convenience. Those characterizations show their natures clearly, and those natures serve different purposes in the narrative arc.

Pris has a brash, aggressive personality that underlies all her interactions, even when she plays coy and shy with Sebastian. While she has perfect control of her body, and uses it to good effect to run away from Sebastian, startle and shock him by sticking her hand into boiling water, attack Deckard, or curl herself sinuously around Batty, she has less control over her personality. The two do not mesh—she is a dualistic being, with her personality being constantly attuned to personal survival, not to seeming as human as possible, even though this would, seemingly, assist that survival. This metaphysical break is evident in how callously she treats Sebastian, as shown earlier. This mirrors the androids of the novella, but does not mirror the movie's version of Rachael. Rachael's body and mind are more fully integrated. Her actions do not work toward her own survival either, but this shows how very 'human' she is.

Rachael does two things in particular which indicate this integration: she cries, and she lives. While it is implied that Batty also cries during his "tears in the rain" (Scott, 1982) monologue, Rachael's tears are spontaneous and due to an existential crisis, rather than knowledge of imminent death. When Deckard tells her that her memories are implants, she starts to cry, slowly and unsuppressedly. Her very humanness betrays her emotions in this instance, showing complexity and nuance. In her tears, the viewer can read betrayal of trust, longing, understanding, and grief for what she thought she was. This, then, underscores to the viewer how different she is from Pris.

Pris' emotions are mere pretense, and we, the viewers, know this. Pris starts out the movie with knowledge of her inner self—she is a Replicant, and she does not expect complexity from herself or her emotions. Her only emotion is to stay alive, and to use any means to do so. And yet, even in this, she fails, as Deckard shoots her dead (Scott, 1982). However, before that occurs, Batty visits Tyrell to discuss the topic of Nexus-6 death. Tyrell tells him that their death is inevitable; he cannot change it.

> "To make an alteration in the…evolvement of an organic life system is fatal. A coding sequence cannot be revised once it's been established."
>
> "Why not?"

"Because by the second day of incubation, any cells that have undergone reversion mutations give rise to revertant colonies like rats leaving a sinking ship. Then the ship sinks.... You were made as well as we could make you."

"But not to last."

"The light that burns twice as bright burns half as long," (Scott, 1982).[4]

Yet, death is where the Replicants most closely resemble their human counterparts. Human beings also inevitably die, whereas the sex robots of today may wear out, but if parts can be replaced or modifications made, they will not die.

Death as Representative of Human Likeness and Humanity

The Latin phrase *memento mori* translates to "remember you will die." While this can be seen as morbid fascination, if we lose sight of the inevitability of death, then death has more of a tendency to be intensely traumatic. To have *memento mori* is to value the time we are here, and use that time wisely, knowing that life's nature is finite. Our knowledge of our own mortality spurs creativity, the urge to produce art, to produce music, to write books, to have children—all things that will outlast our own natural life span. Rosen, Tyrell, and Wallace all refer to their created beings as their children (Dick, 1968/2017; Scott, 1982; Villeneuve, 2017). To bear children is in a way to seek immortality—to accept one's own inevitable death while attempting to ensure that one's own contribution to the world continues after we are gone. Creation implies eventual destruction, and birth implies eventual death.

Thus, both these impulses, to create and to reproduce, tie into death as a representative of the human condition. Kübler-Ross' five stages of grief (denial, anger, bargaining, depression, and acceptance) (Kübler-Ross, 1969) are a reflection of our knowledge of the inevitability of death and the trauma that death carries with it. Denial slowly morphs to acceptance, as we *memento mori*. The Nexus-6 Replicants know this too, their slave-existence being a consistent reminder that they will die and that they are expendable. They do not have the option to create, or even to self-replicate. They have all the impulse without any of the ability. However, they also have considerably fewer years than humans. Does this add or detract from their compatibility on Levy's ten factors?

While it increases one (similarity), it decreases several (desirable characteristic of other, reciprocal liking, filling needs). We have come to expect the things we create to have finite lives—cars break down, appliances just break, and everything mechanical eventually needs either repair or replacing. We try to preserve our creations, and ourselves, through modern medical interventions, but we eventually accept the inevitable; that everything has a finite existence. However, we are unable to completely parse this when it comes to robots. As a non-specialist public, we, collectively, are unaware of the exact parameters of robot 'life' or what it means to have a robot die.

This became evident recently, with the 'death' of the Mars Rover Opportunity, an event that was made official by NASA on February 13, 2019. Opportunity was a data-collecting robot sent to the Martian surface to observe, analyze, and transmit information back to earth. After several months of silence, NASA declared Opportunity 'dead,' sang it to rest with Billie Holliday's "I'll Be Seeing You," and released its last words "my battery is low, and it's getting dark" (ABC7, 2019). The original mission was meant to last 90 days, and the robot persisted for 14 years. In honor of that, and with characteristic anthropomorphizing of our robot creations, we mourned Opportunity as though it had been human. Its poignant and touching last words did nothing to dispel this mindset, being very human, and very alone.

In fact, our mourning of Opportunity closely matches Batty's last words, and their impact on the viewer. "My battery is low, and it's getting dark" (ABC7, 2019) implies the same level of cognizance of death as "All those moments will be lost in time, like tears in rain. Time to die" (Scott, 1982). Yet, the cognizance may merely be *implied* rather than actual, in the case of the Mars Rover. Opportunity almost certainly did not transmit actual words, but rather a data stream reporting on its operational status: a low battery signal, a set of less than optimal light readings from its sensors. The general public would not be able to interpret those data, nor would we be able to understand Batty's internal state when his head falls forward, and his hands open, releasing a white dove into the pouring rain. What we can understand is the similarity that is death—we are born or created. We live, briefly or long. We die when our heart no longer pumps blood, when our eyes no longer find focus and cloud over, when our brain waves are stilled. If those anatomical features, heart, blood, brain, eyes, are mechanical rather than flesh, we can still find commonality and understand that death as an occurrence.

However, we must come back in our discussion to the factors of attraction which death decreases: desirable characteristic of other, reciprocal liking, and filling needs. Inevitably, the focus of these factors of attraction is on the human's attraction to the robot, although in Chap. 4, I will focus on what factors the human provides as well. The death of the mechanical entities in our immediate surroundings often inspires not mourning, but frustration. We thump a device that is not working properly. We get rid of the formerly active husk of a toaster, or a blender, or a television, rather than trying to figure out what went wrong. In instances where we do try to fight the inevitable, such as with a car or a computer, it is often because the machine is valuable (or expensive, which is not necessarily the same thing) and necessary to our continued well-being and our progress through our daily lives. We have created these things, and damn it, they ought to work properly and consistently! Death is not a desirable characteristic of our robots.

Death also lowers the next two—reciprocal liking and filling needs. Machines that break down, or die, are often anthropomorphically perceived as having done so out of spite. Humans can be found to utter phrases similar to 'I hate my computer, it is always giving me the Blue Screen of Death,' or 'my car is acting weird today, maybe I should threaten it with a trip to the mechanic.' When something has gone wrong with our machines, we may also give up. In the clubrealdoll.com forum for Realbotix Harmony users (Clubrealdoll, 2019a), a user named 'User 1'[5] reports that his app froze on install with the generic avatar, and he could not get it to load properly. He emailed the tech support, who had not gotten back to him almost a week after his initial problem, and then expressed that "[S]till end up with a blank screen, no matter if I wait a half hour. Maybe give up if [tech support] does not get back to me" (ibid.).

Other user problems involved lack of a microphone control button in the app depending on the phone substrate they were using, and the app glitching when they asked their doll an un-parseable sentence, or one to which the doll did not have a readily accessible programmed answer (Clubrealdoll, 2019b). It is this last issue that demonstrates needs going unfilled. The users want conversation, and a broken bot or a broken app will disappoint and drive down desire. As much as we understand death, we do not wish to deal with it in our sexual and romantic companions. This may change, however, as sex robots become more normalized into society, and we see them as life partners rather than as useable and disposable slave-like creations.

As previously discussed, evidence suggests that using sex dolls or sexbots does not meet the qualifications for objectum sexuality, but it can be seen as a paraphilia in many cases, particularly when the human is unwilling or unable to engage in sexual activity with humans. Paraphilia being, at its heart, about fetishization, the doll or bot must demonstrate certain characteristics. Perceived life and vitality, or at least the absence of obvious lack thereof, becomes one of those characteristics—for the most part, the paraphilia of sex dolls does not overlap with necrophilia (desire for sex with the dead) or somnophilia (desire for sex with sleeping people).

At the heart of sex doll and sex-bot use is the desire for human-like companionship, shown by how the owners discuss their dolls when things are going well. "User 2" says:

> I'm pretty sure that's what Matt McMullans [sic] intentions are with the Harmony project, is to create a companion that people can talk to and engage in meaningful conversation with. To have something [sic] that cares about you, askes you how your day was, remembers your birthday, and just adds some light to your life when the world can be so dark for so many people, [Clubrealdoll, 2019c].

"User 3" replies "And it's not really about the sex. You can turn that persona aspect completely off, and still have a super companionable experience!" (ibid.). "User 2" then responds by sharing a picture of himself with his companion "Kristal," watching a movie and sharing a bowl of popcorn (ibid.). Yet, as seen in the previous sections, other fetishes and human qualities are not absent from either doll/bot users or the sci-fi that inspired these developments. These include the racialization of the characters in *Blade Runner* (Scott, 1982) and the racial categories available to those who purchase Sinthetics and RealDoll products.

Race and Racialization: Why Does LA Look like Japan?

The aesthetic of the *Blade Runner* universe relies heavily on a visual style which it defined for the succeeding generations of sci-fi movies, which is now called 'techno-orientalism' (Roh, Huang, & Niu, 2015). This is both a nod to the cyber-punk futurism/post-apocalyptic/dystopic sci-fi genres, and to Edward Said's *Orientalism* (1978). The first use of techno, used to describe the music which took that moniker (also sometimes written as tekno), came after *Blade Runner*, in the late 1980s. Before that, the music genre was called 'electro,' and both it and techno are characterized by electronic/synthetic beats, voices passed through vocoder modifiers, and an implied world behind the music that was populated by modified humans, cyborgs, and a sense of a different reality behind what could actually be seen.

Blade Runner (Scott, 1982) emulates this world. However, there is an added layer of orientalism as a running visual and auditory theme. Said's *Orientalism* (1978) explores how certain Western cultures (predominantly the USA and UK), adopt, appropriate, exoticize, and exploit the cultural themes and outward trappings of certain Eastern cultures (India, China, and Japan, though in modern sci-fi such as *Blade Runner 2049* (Villeneuve, 2017), *Cloud Atlas* (Wachowski, Tykwer, & Wachowski, 2012), and others, this can be expanded to Korea as well). Furthermore, the more modern sci-fi gains immense influences from anime/Japanimation, though that cannot truly be argued for the original *Blade Runner* (Scott, 1982) as anime only rose to popularity in the USA within the last two decades, well after the 1980s. What can be argued is that the massive flow of technological information between the USA and Japan during the 1970s and 1980s gave rise to concerns that Japan was going to 'buy' the USA, or at least significant portions of it (NAP, 1997, pp. 26–44). These concerns can be seen in the world-building of Scott's *Blade Runner* (1982) aesthetic.

Set in 2019 LA, the streets and buildings in *Blade Runner* (Scott, 1982) are layered with Asian influences. Chinese character-writing and Kanji emblazoned in neon and paint direct people to restaurants and shops. Massive hologrammatic or light-display-based Geisha figures advertise products on the sides of buildings. The shops are open-sided, with hanging lanterns adorning them and wares displayed as though in an open-air market. This was Scott's own addition to the story—the cityspeak which includes parts of various Asian languages, and the Asianized aesthetic of the people, do not occur in the novella. And while Scott was almost certainly referencing the fear of Japanese takeover of American technology, he also participated in the enhanced exoticizing and fetishizing of the 'oriental' aesthetic in American popular culture.

However, that aesthetic does not extend to the main characters. Although certain supporting characters are Asianized (Hannibal Chew, the eye manufacturer; Howie Lee, the sushi chef; and Abdul Ben Hassan, the snake maker), none of the cops or the Replicants bear any trace of stereotypically Asian features, nor do they have accents or dress in Asian clothing. While Prater and Fung (Roh et al., 2015) make an argument, based on Larissa Lai's poem analyzing the character of Rachael, that Rachael can be read as either Asianized or as passing-mixed-race, as can J.F. Sebastian, this argument both ignores and misinterprets various aspects of the source material.

The elements which are ignored and misinterpreted are Rachael's Jewishness, the Voight-Kampff questions having been taken directly from the novella, and the connection between the novella's J.R. Isidore and the movie's J.F. Sebastian. While I have already addressed Rachael's Jewishness from the point of view of the novella, it does carry over in one significant way in the movie: her hair. The movie's incarnation of Rachael usually presents as nothing short of immaculate, with her hair in a very classic style—a French knot swept up at the nape of her neck, and eyebrow-length bangs, carefully curled and shaped to frame the rest of her face. However, when she surrenders to the knowledge of her Replicant nature, at Deckard's apartment when he wakes while she plays the piano, she takes her hair down and reveals that it is a riot of waves and curls, untamed and extraordinarily stereotypically Jewish, particularly in depictions from the 1980s (Tablet, 2015). If one had not read *Do Androids Dream of Electric Sheep?* (Dick, 1968/2017), or had not understood the significance of both 'Rachael' and 'Rosen' as name choices, one might code this as 'mulatto'[6] as Prater and Fung do, but it is more properly read as a continuation of Jewishness as 'the other.'

The second aspect is that of the use of the Voight-Kampff question regarding the consumption of both raw oysters and dog meat. Prater and Fung interpret this as anti-Asian, stating "[T]he question that reveals Rachael to be a Replicant invokes the Western perception of Asians as dog eaters" (Roh et al., 2015, p. 199). However, this question comes directly from the novella unchanged, and is more properly read to show Rachael as, again, Jewish—in various sects of Judaism, dogs are one of the non-kosher, unclean creatures, as are shellfish of any kind. Both dogs and oysters are forbidden as food, and dogs are proscribed as pets. While correctly pointing out that 1982 was not temporally distant from desegregation and that America (and our science fiction) was still in the process of working out racial issues, one need not appeal to the black/white dichotomy for an explanation of the character of Rachael, when her Judaism is canonically accurate and explicitly signaled.

The final point is that of the race/racial features of J.F. Sebastian. Prater and Fung indicate that Sebastian belongs to a long history of encoded racial features: "Sebastian becomes readable in the genealogy of 'tinkerers' who created and sustained a network of relations of colonial capital and the discourses of race, sex, sexuality and gender that undergird the institutions of slavery and imperialism" (ibid., p. 196). However, the authors overreach when attempting to connect that to the other roles played by the actor. Typecasting is just as meaningless as stereotyping, in this particular case, because the character of Sebastian is not meant to mimic "southern 'hicks' and 'moonshiners'" (ibid.) which the actor previously played, but to reflect the chickenheads seen in the novella.

While Scott codified the degeneracy seen in the novella into the genetic condition "Methuselah Syndrome" (Scott, 1982), the premise remains the same—those who are defective, those who are poor, those who are unable or unqualified to leave the planet are stuck on earth as it succumbs to the contamination (presumably nuclear) from whatever war occurred. The on-world/off-world dichotomy is the same. Replicants are allowed to exist only off-world, and they do the dirty work that humans do not want to do. Humans are allowed in either place, but only the perfect humans, where perfect is coded as rich, healthy, attractive, white, are allowed to move off-world and experience a life away from contamination. However, in this case, the so-called perfection of whiteness is only contrasted with Asian, and there are no Asians in disguise and no mixed-race individuals passing for white in this world, as was hypothesized by Prater and Fung (2015).

However, where one does see the intense exoticization of multiple various racial groups is in the construction of sex dolls and sex-bots. From names to skin tones (and the names of those skin tones), the dolls and bots are highly exotic and very racialized. From the Sinthetics website, the skin tone descriptions are as follows:

Cream (fair) – Northern European, Asian, fair Latin
French Vanilla (medium fair) – Caucasian, Asian
Peaches (pinkish fair) – Caucasian such as Northern European, Scottish etc
Honey (golden fair) – Caucasian, Asian, Latin
Sugar (pinkish medium) – Caucasian medium skin tone to replace Latte
Chai (light olive) – Asian, Southern European, Latin
Spice (pinkish tanned) – Caucasian, Latin, Southern European to replace Caramel.
Dolce (medium tanned) – Latin, Asian, tanned Caucasian, Indian, Pacific Islander, mixed ethnicity, Native American
Toffee (dark tanned) – Latin, Asian, mixed ethnicity, Native American, Pacific Islander, Indian

Mocha (deep tan) – African, Latin, mixed ethnicity, Native American, Indian, Middle Eastern
Cocoa (deep brown) – African, Middle Eastern, East Indian
Chocolate (deeper golden brown) – African, Middle Eastern, East Indian
Espresso (deepest brown) – African, East Indian [Sinthetics, 2012–2019].

All of these can be paired with the following head types (photos of which can be found on the website): Alicia, Belinda, Celeste, Celestine, Eliza, Kimiko, Monique, Tawny, Yuriko, and Willow. Of these, the first five plus Tawny have Anglo/white features, while Kimiko and Yuriko are both Asian, Monique is black, and Willow is an elf. Ignoring the elf, for the moment, we can see that the racialized facial features and corresponding default skin tones of Kimiko, Yuriko, and Monique are matched to their names. The company relies on multiple visual and verbal cues to signify race—including the fact that all of the skin tones are food-based, which is a racialized signifier—ensuring that the customer can choose whatever doll matches their desire or paraphilia.

Similarly, RealDoll uses their own naming conventions to do likewise. While there are only five skin tones (cocoa, fair, light tan, medium, and tan), there are 37 standard faces, seven of which are photorealistic doubles for living people—pornography actresses Alektra, Asa Akira (Asian), Jessica, Kaylani (Latina), Lupe (Latina), Samantha, and Stormy Daniels of the Donald Trump presidency scandal fame. Those who are not pornography star doubles still often have names that match their racial features. Looking at only the non-white featured heads, we see Amara (black), Sai (Asian), Nova (Latina), Solana (Latina), and Violet (black).

The ability to customize does not undermine the deliberate racialization, nor the fact that the racial features used in the facial constructions are both stereotypes and exoticized to appeal to sexual fetishes. Regarding one of the pornography star doubles, an anonymous reviewer writes:

> My Lupe (WRD-Lupe) was a dancer on the Descontrol TV show. This one is the best representative of a Latina pop star. She is very small, really cute, and physically fit. Except for her excellent breast size she could be a gymnast. They have shaped the body perfectly for a womean [sic] this size. XS clothes are a bit big on her. The makeup is a natural look. Compared to the doll the internet tie-in is only marginally interesting. This one is easy to pose, so you never know where in the house she will turn up. I can't overly stress the cutness [sic] factor, [RealDoll, n.d.].

This idea of the dolls and bots as representatives of real people, or of real archetypes, stems directly from the personalization of the Replicants in *Blade Runner* (Scott, 1982). And just as medical robotics and cyborg technology was driven by Dick's novella (1968/2017), so too did the industry surge after the 1982 movie.

Developments in Robotics: 1980–2000—Cognition, Neuromorphic Chips, and Biofeedback

Much of the developments in computational power in the 1980s and 1990s centered around the idea of the thinking computer, focusing on how humans think and designing computers to work in similar ways. With AI seen as a promising development for everything, from toys to medical enhancements, the scientific community's efforts centered around the dual questions of what cognition is and what computers were capable of in terms of it (Waldrop, 1988). This first question remains unanswered—we still do not completely understand human cognition and the consciousness that is required for it. The brain, long acknowledged as the seat of cognition and consciousness, has the ability to compensate for injury or illness, and while we are edging closer to understanding such things, including studying how individual neurons interact with each other to form neural networks and specialized areas, we are far from definitive answers to these questions. It should also be noted that computers, thus far, do not have this compensatory ability—they remain non-adaptive enough that their point-of-no-return for damage is at a considerably lower threshold than that of the human brain, and that the propensity for both software and hardware to develop unreconcilable glitches over time results in the need for frequent replacement. This, then, creates a recursive problem for cyborg/cybernetic scientists: how to recreate/create something like the human brain or human cognition, if we do not have exact computational parameters or hardware specs.

The science of creating artificial cognition began with the basic knowledge that neurons interconnect via dendrites and axons, forming dense, adaptational networks of cells. Each cell is in communication with several others, allowing for multiple patterns of informational combinations to occur as needed to produce the desired output. These networks, in combination with various neurotransmitters, create varying excitation levels which we experience as feelings, thoughts, words, actions, sight, and every other thing that constitutes cognition and consciousness. Through MRI (magnetic resonance imaging, which is used to take high-resolution magnetic scans of organic tissue) and fMRI (functional MRI, which creates function-patterns of brain activity), we can see not only the brain, but visual representations of the thoughts and emotions of the person being studied. However, much of this scientific ability to see such things has happened only recently, post-2000s, and seeing does not translate to recreating.

This idea of the neural network led to the creation of neuromorphic computer chips. In the late-1980s/early-1990s, Carver Mead, of Caltech, did extensive work to create a chip that would behave like a neural network (Mead, 1990). His premise was that in order for computers to instantiate the dynamic structure of the human mind, to approximate the complexities of a neural network, they would have to physically mimic the brain via a series of adaptational, variably excitable, synaptic-like connections. Mead identified various problems (ibid., p. 1630) such as size, weight, electric power requirements, and the various costs associated with all of these. A working model of a brain is no good if you need to go to it, rather than taking it with you, and at this time, the World Wide Web was in its infancy as a commercial entity, being only a few years past governmental and military embargo.

Initial examination of the capability of brain-like computer processors included ideas of emotional and evaluative authenticity, as well as computational power. One such example is the concept of art. The differentiation, in this case, was between rational computation and intuitive and emotional creation (Maiocchi, 1991). This is incredibly similar to the distinction made between the Replicant's ability to mimic human behavior, but only when that behavior was self-preservation based, that is, logical. They could not create their own personalities, as a personality is emotion-based, irrational, and intuitive. In fact, the hardware/software/output issue became vitally important in this particular experiment, with Maiocchi stating that one must determine "if the art resides in the program that generates the piece or in the output itself" (p. 185).

Various concepts in this experiment translated badly to a strictly rational computer program, where rational means that all outputs are a product of programming plus hardware capacity plus instructional input from the user or programmer. In this case, such abstractions such as beauty, which he defines as containing "qualities of intelligence such as learning, creativity, emotional responses, memory, and a sense of self" (p. 186), were difficult to design into a program, and even more difficult for the program to implement. However, such initial explorations laid the groundwork for predictive text (colloquially known as autocorrect), emotional recognition software, and empathetic AI, all of which will be discussed further in subsequent chapters.

Also in its beginning stages, during the post-*Blade Runner* time period, was the idea of biofeedback and its use in medical interventions. While this concept would not merge with cybernetics for quite some time, it is worth exploring the origins to understand how its use in prostheses came to be. A quick survey of the literature being produced during this time shows us that the focus of biofeedback—done, necessarily, with external electrodes and monitors—was in allowing patients to become knowledgeable about their own bodily functions so that they could attempt to assert control over them. Examples such as the possibility of lowering one's blood pressure or heart rate, relieving muscle tension or headache, and triggering or suppressing nerve impulses were all foci of research during this time (Pikoff, 1981).

This research concentrates not on feedback from an artificial device to the brain, as we will see in later technology, but from the brain to itself via a mediator, such as those aforementioned electrodes and monitors. The body, naturally, gives feedback at all times, through pain, pressure, temperature, and proprioception nerves. This type of biofeedback explains why we pull back after touching a hot stove, can walk (mostly) unaided down a set of stairs, have the tactile awareness to stroke our pets gently instead of hitting them, and in general have an understanding of what happens when we injure ourselves. We constantly, and subconsciously, monitor, respond to, and regulate our actions to account for the feedback given to us by our own bodies. Such feedback was largely absent in the Replicants, as demonstrated by Batty's superhuman abilities to punch through walls and leap unfathomable distances, and Pris' ability to stick her hand in boiling water in order to horrify Sebastian. It is that moderating word, 'subconsciously,' that such biofeedback research sought to change, and the ways which this research changed the technology, and the stories we tell ourselves, will be discussed further in the next chapter.

Notes

1. I have chosen not to quote the movie in this particular instance, so as not to write that word out in this text.
2. Also a reference to the chickenheads of the book.
3. The question of biological waste is not addressed, as it often is not in movies with a certain amount of seriousness to them; however, one presumes that Replicants (and the Androids before them) do eliminate in similar manners as humans.

4. This language is mirrored by Villeneuve in *Blade Runner 2049*, when Joi laments that her information only contains two symbols (1 and 0) to K's four (ATCG). K replies "half as much, but twice as elegant, sweetheart," (Villeneuve, 2017).
5. User names removed to respect anonymity.
6. It should be noted here that, at least in anthropology, this word is no longer considered accurate or polite, and is only to be used in contexts where one is describing or transcribing an historical usage. The proper anthropological term as of 2019 is biracial, mixed race, or mixed ancestry.

Bibliography

ABC7. (2019). My battery is low and it's getting dark. Retrieved from https://abc7chicago.com/science/my-battery-is-low-and-its-getting-dark-opportunitys-last-message-to-scientists/5137455/

Campbell, N. (2010). Future sex: Cyborg bodies and the politics of meaning. *Project Muse, 11*(1). https://muse.jhu.edu/article/377515/summary

Clubrealdoll. (2019a). Retrieved from http://clubrealdoll.com/community/threads/realdoll-x-stuck.2779/

Clubrealdoll. (2019b). Retrieved from http://clubrealdoll.com/community/forums/harmony-app-bug-reports.25/

Clubrealdoll. (2019c). Retrieved from http://clubrealdoll.com/community/threads/harmony-what-potential-do-you-think-is-there.2663/

Danaher, J., & McArthur, N. (2017). *Robot sex: Social and ethical implications*. Cambridge, MA: Massachusetts Institute of Technology.

Dick, P. K. (1968/2017). *Do androids dream of electric sheep?* New York, NY: Del Ray.

Forbes, B. (Director), & Scherick, E. J. (Producer). (1975). *The Stepford Wives* [Motion picture]. USA: Palomar Pictures.

imdb.com. (2016). Artificial females in movies and television. Retrieved from https://www.imdb.com/list/ls063279126/

Kübler-Ross, E. (1969). *On death and dying*. New York, NY: Macmillan.

Levin, I. (1972). *The Stepford wives*. New York, NY: Random House.

Levy, D. (2007/2008). *Love and sex with robots: The evolution of human-robot relationships*. New York, NY: Harper Perennial.

Maiocchi, R. (1991). Can you make a computer understand and produce art? *AI and Society*. London, UK: Springer-Verlag.

Mead, C. (1990). Neuromorphic electronic systems. *Proceedings of the IEEE, 78*(10), 1629–1636.

Menzel, P., & D'Aluisio, F. (2000). *Robo sapiens: Evolution of a new species*. Cambridge, MA: Massachusetts Institute of Technology.

NAP. (1997). https://www.nap.edu/read/5850/chapter/5

Oz, F. (Director), & Rudin, S., DeLine, D., Scherick, E. J., & Grunfeld, G. (Producers). (2004). *The Stepford Wives* [Motion picture]. USA: DeLine Pictures.

Pikoff, H. (1981). Biofeedback: A resource directory and outline of the literature. *Professional Psychology, 12*(2), 261–270.

Prater, T., & Fung, C. (2015). "How does it not know what it is?": The techno-Orientalized body in Ridley Scott's *Blade Runner* and Larissa Lai's *Automaton Biographies*. In D. S. Roh, B. Huang, & G. A. Niu (Eds.), *Techno orientalism: Imagining Asia in speculative fiction, history, and media* (pp. 193–208). New Brunswick, NJ: Rutgers University Press.

RealDoll. (n.d.). Home page, testimonials, and product information. Retrieved from realdoll.com

Roddenberry, G. (Producer). (1966–1969). *Star Trek: The Original Series* [Television series]. Hollywood, CA: Desilu Productions.

Roh, D., Huang, B., & Niu, G. (2015). Desiring machines, repellant subjects: A conclusion. In D. S. Roh, B. Huang, & G. A. Niu (Eds.), *Techno orientalism: Imagining Asia in speculative fiction, history, and media* (pp. 221–226). New Brunswick, NJ: Rutgers University Press.

Said, E. W. (1978). *Orientalism*. New York, NY: Pantheon Books.

Scott, R. (Director), & Deeley, M. (Producer). (1982). *Blade Runner* [Motion picture]. USA: The Ladd Company.

Sinthetics. (2012–2019). Testimonials and product descriptions. Retrieved from http://sinthetics.com/

Spielberg, S. (Director), & Kennedy, K., Spielberg, S., & Curtis, B. (Producers). (2001). *A.I Artificial Intelligence* [Motion picture]. USA: Amblin Entertainment.

Swift, D. (Director), & Disney, W., & Golitzen, G. (Producers). (1961). *The Parent Trap* [Motion picture]. USA: Walt Disney Productions.

Tablet. (2015). The art of Jewish hair. Retrieved from https://www.tabletmag.com/jewish-life-and-religion/190600/the-art-of-jewish-hair

Villeneuve, D. (Director), & Kosove, A. A., Johnson, B., Yorkin, B., & Yorkin, C. S. (Producers). (2017). *Blade Runner 2049* [Motion picture]. USA: Alcon Entertainment.

Voskuhl, A. (2013). *Androids in the enlightenment: Mechanics, artisans, and cultures of the self*. Chicago, IL: University of Chicago Press.

Wachowski, L., Tykwer, T., & Wachowski, L. (Directors), & Arndt, S., Hill, G., Wachowski, L., Tykwer, T., & Wachowski, L. (2012). (Producers). *Cloud Atlas* [Motion picture]. USA: Cloud Atlas Production.

Waldrop, M. (1988). Toward a unified theory of cognition. *Science, 241*(4861), 27–29.

Wosk, J. (2015). *My fair ladies: Female robots, androids, and other artificial eves*. New Brunswick, NJ: Rutgers University Press.

CHAPTER 4

Angel Replicants and Solid Holograms: *Blade Runner 2049* and Its Impact on Robotics

Abstract In this final descriptive chapter, Gibson examines the most current iteration of the story, *Blade Runner 2049*, and teases out implications for the future of robotics. Humans identify themselves against a contrast of Replicant life in a very binary fashion. The creator is meaningless without the creation, for to know what it is to be human, there must be an inhuman correlate. For the latter part of this chapter, the author includes interviews with robotics developers and sci-fi authors about their own desires, and why they chose to make hominoid robots. What happens when the creator is replaced and the creation can give birth to life? Will our own robotics ever be able to self-replicate?

Keywords *Blade Runner 2049* • Replicants • Sex • Luv/love • Joi/joy • Reproduction

Synopsis

In *Blade Runner 2049*, Denis Villeneuve's (2017a) sequel to Ridley Scott's *Blade Runner* (1982), we are introduced to the newest model Replicants—these Replicants, built after a complete overhaul of the business, are made to obey. They do not step outside their programming, they do not rebel, so they can be tasked with performing jobs no one else wants without the constant possibility of violence.

R. Gibson, *Desire in the Age of Robots and AI*,
https://doi.org/10.1007/978-3-030-24017-2_4

Personifying these new models, police officer KD6-3.7 (spoken as K D six dash three dot seven), played by Ryan Gosling, has been assigned as a Blade Runner. While being a Blade Runner was prestigious in the original movie, it has lost respect and is now a job given to the Replicants, who are not respected at all. The movie opens with K, as he is called, retiring an older model—a Nexus-8, who had been in hiding. The Nexus-8, Sapper Morton, played by Dave Bautista, farms grubs for protein, and has been quietly minding his own business since the rebellion; however, he attacks K who beats him into submission before killing (retiring) him, removing his eye, and bringing the eye back to his car for identification with his superior, Lieutenant Joshi (played by Robin Wright). While there, he spots a small sprig of primroses at the base of a dead tree—he asks his flying camera to investigate, and with x-ray tech, the camera spots a buried box below the ground. Joshi sends a dig team and tells K to return to the office (now shaped like a huge nail—more on this symbolism to come) for his 'baseline' test.

The baseline has replaced the Voight-Kampff test. All Replicants are now registered and tagged with a serial number, and they have individualized baseline tests which indicate that they are in working order. K's test is built around a passage from Pale Fire by Vladimir Nabokov, but quickly takes a not-so-subtle turn into taunting him for his status as a Replicant. The original passage is:

> And blood-black nothingness began to spin
> A system of cells interlinked within
> Cells interlinked within cells interlinked
> Within one stem. And dreadfully distinct
> Against the dark, a tall white fountain played [Nabokov, 1962/1989, p. 59].

However, the interviewer expands on the theme, asking him provoking questions about isolation, solitude, and the (lack of) human connection indicated by being a Replicant. K answers calmly and correctly, looking into the lens which verifies his identity and rank.

Having passed his baseline, K goes home to wash off the blood, treat his injuries, and have dinner. When just inside his apartment, he triggers a wall console which plays a short tune, and then speaks to someone off-screen. The woman answers back, telling him she is almost done cooking dinner, telling him to go shower. He says he needs a drink, does she want one? She does—he pours two drinks, and then drinks them both. As she comes in with 'dinner,' we see why—she is an anthropomorphic hologram, being projected by the wall console and a moving mechanical arm that swivels as she moves around the apartment. This is Joi, played by Ana de Armas. Joi, a companion hologram, can be customized to suit "everything you want to see, everything you want to hear" (Villeneuve, 2017a). K chose to set her up with these stats:

Height	168 cm
Body type	Slender
Face type	Classic
Skin tone	Olive
Eye color	Brown
Lip color	Neutral
Hair color	Brunette
Hair style	Bangs
Ethnicity	Cuban
Language	English [ibid.].

For a surprise, K has bought an emanator, a mobile console that allows Joi to manifest as *almost* tangible—she can 'feel' very slightly and touch very slightly, though she is still far from solid. And she can travel with K, in his pocket, manifesting seemingly at will.

Just as K leans in to kiss the newly semi-solid Joi, he gets a message from Joshi that overrides the Joi module, calling him back to work—the extraction team has found the box from Sapper's farm, and it has a skeleton inside it. The skeleton has nicks on the pelvis, which the examiners take to indicate a botched C-section.[1] Zooming in further on the pelvis, K notices numbers; a serial number. Joshi immediately realizes what this means. The woman, the skeleton, was a Replicant. And she was pregnant.

Joshi and K retreat to her office, where she orders him to find out everything he can about the Replicant, find the child, and take care of everything, destroying it. He goes to the Wallace Corporation, which replaced the Tyrell Corporation as the producer of Replicants after the business overhaul. Headed by Niander Wallace and staffed with a combination of humans and Replicants, the Wallace Corporation has all the records for older models/serial numbers. K gives over a lock of hair from the box of bones, and the serial number comes up as a model that was the "last gen, pre-prohibition" (ibid.); however, the file clerk cannot find more information. He hands the card off to Luv, a striking female Replicant played by Sylvia Hoeks, who brings K to a more secret storage room, deep inside the corporation. During their walk, the Joi emanator module rings, and Luv remarks, "I see you're also a customer. Are you satisfied with our product?" K replies, "She's…very realistic, thank you" (ibid.). The corporation building still retains the form of a ziggurat, and every part of it shows how wealthy and powerful Wallace is—everything solid is made of wood, every room has water and sunlight in it, which deeply contrasts with the rest of the society, as shown by K having only a five-second burst of water for his shower.

Luv finds an optical data sphere regarding the old serial number, and it shows part of Rachael's Voight-Kampff interview with Deckard. The bones are of Rachael's, and she had been pregnant. K leaves and seeks out Deckard's old partner, Gaff. Gaff is unhelpful, telling K only that Deckard is retired, while folding an origami sheep, a callback to the novella (Dick, 1968/2017). The movie returns to the Wallace Corporation, where Luv brings Niander Wallace (played by Jared Leto) to see the newest model Replicant. Blind, Wallace sees by way of neural implants and small flying video bots that the implants control—he has only to think of where they should position themselves, and they move to 'see' for him. The new model is female, but sterile, like all the rest. Luv, though she would do anything for Wallace, cannot fix this problem, and she is clearly terrified of his anger. She watches, barely reacting, as he caresses the Replicant, and slits her abdomen like he is gutting a fish, then kisses her lips as she drops to the ground. The Replicant dies as he walks away, ordering Luv to bring him Rachael's child. As he leaves, he calls Luv the best angel of all he has made.

K stops at a food court for lunch. We see a pimp tell three of her Replicant prostitutes to go learn what K knows about Sapper Morton's death, but when they find out that he's a Blade Runner, two of them opt to leave him alone. The third, Mariette—played by Mackenzie Davis—tries to chat him up, but the Joi emanator module rings in his pocket, and she comments "Ohhhh…you don't like real girls. Well. I'm always here" (ibid.). He goes back to Sapper's farm, and discovers three things—a baby sock, a photo of a woman and a child in front of the tree, and, inscribed on one of the tree's roots, a date: 6/10/21. Joshi gets notification that Luv has stolen Rachael's bones and murdered one of the morgue attendants, and we cut to the interior of K's apartment. K tells Joshi a memory, from when he was 'little.' It is an implant, but it is personal, and they are sharing a moment of intimacy outside of the case. She is drinking his vodka, getting more and more drunk, and more and more flirtatious. Eventually, she propositions him, and he sidesteps by saying that he should get back to work.

His next stop is the DNA records room, to look for children born on 6/10/21. Joi helps him look, and calls him out on the fact that he lied to Joshi—he did not tell her that the date 6/10/21 is connected with his childhood memory. However, they are being watched, so he shushes her musings on whether or not *he* is Rachael's child. Together, they find two identical DNA records, both born on 6/10/21, one boy and one girl. This, of course, is impossible genetically, particularly as the girl was said to have died of "Galatian's Syndrome" (ibid.), which would make them not identical after all. K and Joi decide to go to the orphanage where the boy was stashed to find out what happened.

He recognizes it as the same orphanage from his memory, and follows the clues left in the memory to find a small carved wooden horse (really, a unicorn with the horn broken off, a callback to Gaff's origami from the first movie). The records from 2021 have been destroyed, but K now feels even more strongly that he is the child he is looking for. In order to test this, he goes to the memory maker, Dr. Ana Stelline, played by Carla Juri. She is immunocompromised, so she stays locked in a chamber creating memories from her memories. She confirms that K's memory is 'real': "someone lived this, yes" (ibid.). K takes this as confirmation that he is Rachael's child, and reacts very emotionally.

However, K has not communicated with Joshi during this process, and Joshi thinks he has gone rogue—she sends other cops to bring him in. During a subsequent baseline test, his emotional upset shows. He fails the test, but tells Joshi that he has found the child and taken care of the situation. Joshi puts him on leave so that he can get a hold of himself. Returning home, Joi greets him, and there is a knock at the door. It is Mariette. Joi has contacted her, asked her to come have sex with K as Joi syncs into Mariette's body. Joi wants to be solid for K, and this is the only way she can think to do so. When she leaves in the morning, Mariette slips a tracking device into K's coat.

K realizes that he has to run, and Joi insists on coming with him by transferring all her data to her emanator module. When he breaks the main antenna, it alerts Luv to the fact that he is about to flee. Her first move is to check his apartment, and then she walks into Joshi's office and tries to intimidate K's whereabouts out of Joshi. Luv does not succeed and kills Joshi, activating her office computer equipment by holding Joshi's dead face in front of the ID console. She uses the tracking software to follow K's movements.

K's first stop is at a dealer in rare artifacts—the wooden horse/unicorn checks out as being 30 years old, made with radioactive wood from the dirty bomb that caused the worldwide blackout. K uses this information to find Deckard, who lives in a derelict casino in Las Vegas, with a dog. Deckard, having wanted to be left alone, has boobytrapped the entire place, and he and K fight to a standstill before agreeing to have a drink and talk. K learns that Deckard left before Rachael gave birth; he never saw his child. Deckard has carved other animals in the same manner as the horse/unicorn. K asks why he left, and Deckard replies that "we were being hunted. I didn't want our child found, taken apart, dissected. Sometimes to love someone, you've got to be a stranger" (ibid.). During their confrontation, both Luv and Mariette have been tracking K.

Luv's forces attack first, easily knocking out Deckard and strapping him, unconscious, into a waiting transport. K keeps fighting, and Luv deals with him personally, managing to subdue him after he has been shot, stabbed, and hit in the head multiple times. The emanator has fallen out of his pocket, and he reaches for it. Luv sees him do so, and Joi manifests as Luv goes to step on it, saying "I do hope you're satisfied with our product." Joi looks to K and says, "I love you," just as Luv 'kills' her (ibid.). He passes out, and wakes having been bandaged by Mariette. Mariette has been joined by the 'pimp' from earlier, a Replicant named Freysa, who leads the secret Replicant rebellion which has been protecting Rachael's child since birth, trying as well to unlock the method of Replicant reproduction. Freysa dispels K's notions of his own specialness; Rachael's child was a girl.

Deckard wakes in Wallace's lair. Wallace asks for the location of the child, which Deckard does not know. Wallace shows Deckard Rachael's skull, caressing it fondly. He tries to make Deckard question his own humanness:

> 'did it never occur to you that's why you were summoned in the first place? Designed to do nothing short of fall for her, right then and there. All to make that single perfect specimen. That is, IF you were designed. Love, or mathematical precision. Yes, or no.'
>
> 'I know what's real,' [ibid.].

When self-doubt does not work, Wallace turns to emotional blackmail. He has Luv bring out a female Replicant in the exact form, features, and dress of Rachael, the first day Deckard met her. However, the details are wrong,[2] and Deckard rejects her, saying, "her eyes were green"[3] (ibid.). Luv shoots her in the head, and Wallace turns to his next threat: pain. They will take Deckard off-world, and persuade him to talk with torture.

We move to a shot of K walking home after his discussion with the rebellion. He is approached by a huge hologrammatic advertisement for Joi, who addresses him using some of the information that his own personal Joi had in her memory banks before she got destroyed. Dispirited, he knows he could purchase her again, recreate her perfectly, but it would all be fake. In that moment, he decides to go after Luv and Deckard, to help the rebellion in protecting and saving Rachael's child, the child that is not him.

K shoots down the transports that are escorting Luv and Deckard to the airport, and Luv's transport crashes into the shore of the Pacific Ocean. K sets his car down near theirs, and shoots Luv through the glass. The transport begins to fill with water, as Luv and K fight to the death, with Deckard still handcuffed to his seat. Luv eventually stabs K in the side, kisses his bloodied lips, and whispers, "I'm the best one" (ibid.), before trying once more to get the transport to start. K surprisingly recovers enough to strangle and drown her, before rescuing Deckard and flying him off to the memory center to meet his daughter, Dr. Ana Stelline. K dies on the steps of the center, having given Deckard the horse/unicorn—a callback to the final scene of the original movie, when Rachael kicks over Gaff's origami unicorn, which Deckard picks up.

What Is in a Name?: The Functional and Metaphorical Contrast of the Joi/Luv Dichotomy

With 35 years between the release of *Blade Runner* and *Blade Runner 2049*, the characterizations have become more detailed and complex—even those with direct ties back to the novella. Two of the most vital characters, with detailed characterizations, are Joi and Luv, and they are only tangentially connected to *Do Androids Dream of Electric Sheep?* (Dick, 1968/2017) and *Blade Runner* (Scott, 1982). Joi is an anthropomorphic hologram. According to her advertising, seen many places in the movie, she is "everything you want to see…everything you want to hear…everywhere you want to go" (Villeneuve, 2017a). The Joi hologram is completely customizable, as seen by her taglines and her vital statistics, glimpsed when K updates her profile to the emanator unit. For each trait (hair color and style, eye color, height, etc.), there are other choices, perhaps infinite, that the customer can make.

In contrast, Luv also exemplifies everything one person wants, but she does so in a perfectly singular manner—she is the unique production of Niander Wallace, his best angel, and she serves him and him alone. She began as completely customizable, too. Wallace has complete control over his new model Replicants, creating in them the size, shape, appearance, and personalities that will suit himself or his clients, and so we see in Luv the perfect creation—tall, strong, female, obedient, but fiercely intelligent, loyal, willing, and able to do anything, anything at all, in his service.

In this way, the two are functionally opposed. Joi is a pleasure hologram. As she is non-corporeal, she serves only one purpose: companionship. That companionship will look different to each customer. Some will want an electronic nanny. Some, a responsive sex kitten to lounge around their home. Some, a person to flatter them. Some will desire a verbal sparring partner, someone to contradict them, to bounce ideas off. Joi can be all these things, and is, by turns, many of them to K. She flatters, cajoles, cooks for him in a little blue dress with an apron, dances, offers him a book so he can read to her, and offers her opinion on many things in his life, from the book they are reading (Nabokov's Pale Fire (1962/1989)) to whether or not K is Rachael's child. She quite literally adapts to his personality and mood moment to moment, so that she can be everything he wants, exactly as advertised.

Luv's life exists for no one's pleasure. She exists to execute the will and desires of Niander Wallace, and she lives in fear of the potential for disappointing him. Whereas Joi can be switched on or off, and has reactions that include an internal chronometer but no real burden of the passage of time—for example, she 'missed' K while he was at work and was "getting cabin fever" (Villeneuve, 2017a), this is pretty well established to be for show, as it is what K wants to hear—Luv exists in both time and space, and has tasks to perform that create real-world consequences. This can be seen when Wallace comes to approve the newest model Replicant. He asks "[C]an you at least pronounce a child is born?" (ibid.). She stays silent, almost vibrating with fear, crying expressionless silent tears. The new model is presented, and he physically examines her with his hands on her body before Luv hooks up Wallace's eye-bots.

Wallace expresses anger at humans' hypocrisy in using manufactured 'slaves,' the Replicants, when they will not enslave their own. He wishes to create self-replicating Replicants, and mourns:

> [T]hat barren pasture, empty and salted...right here [he cups the new model Replicant's abdomen, where her uterus would be, and caresses her]. The dead space between the stars. And this the seat that we must change for heaven. [He slices the new model across the abdomen with a scalpel]. I cannot breed them. So help me, I have tried.... Tyrell's final trick. Procreation, [ibid.].

He orders Luv to bring Rachael's child to him, and during the entire scene, she cries, knowing that his disappointment will be deadly. However, whereas Joi would adapt to suit any purchaser, Luv's desire to satisfy relates only to Wallace. Her focus remains on him and his wishes throughout the movie.

She serves multiple purposes—she makes sales pitches, arranging to fulfill customer's unique needs; she escorts special visitors, leading K to the information about Rachael and Deckard; and she drives off looters when K goes to visit the orphanage. This last scene speaks of certain parts of her character. After finding anomalies in the DNA database, K and Joi fly to the orphanage to look for the (male) child. Looters attack K's car, firing surface-to-air missiles, one of which must contain some sort of electromagnetic pulse (EMP), because it knocks out all of K's car's functions and knocks out Joi's emanator unit as well. When the car crashes, the looters attack and engage K in hand-to-hand combat. He is winning, but he is completely outnumbered. However, before the looters can overwhelm him, air-to-ground missiles begin falling. We hear Luv's voice say, "Fire again. Fire again. Fire. ...200 feet to the east. Fire. Go north. Fire. Stop. 20 degrees east. Stop. Zoom in. Closer. Oh, come on. Get up. Do your fucking job. Find the child" (ibid.). To target, she wears enhanced reality glasses, seeing a video feed of K's location during the air-assault. She is also having her nails done—a technician with a small laser torch etches holographic designs into each individual nail as she directs the aerial bombardment.

Here we see her complex characterization. While Villeneuve and Hoeks could have worked together to make the character one-sided, flat, and emotionless, Hoeks' brilliant job acting Luv created myriad emotions, many dimensions, with very little expressional changes. Using micro-expressions, minute changes in her tone of voice, and exquisite attention to detail, Hoeks gives Luv an intense inner life. We sense her terror and devotion to Wallace, her own desire to look her absolute best (shown by her nail art, and by her clothing and hair choices throughout the movie), and her disdain for K's imprecision, for Luv is nothing if not precise.

This plays out almost exactly the opposite in Joi, though superficially their characters look very much the same. Tall, slender, pale, dark-eyed, and with medium-length dark hair and bangs, their differences are most evident in the features that evince their personalities. The way they wear their hair, the way they move their bodies, their facial expressions, all show the dichotomous characterization of Joi and Luv. Where Joi has a 'sloppy' bun, Luv's hair is immaculate. Where Joi moves languidly, by turns perky, innocently seductive, casually graceful, and softly loving, Luv's movements are calculated, precise, exactly enough to do what is necessary and nothing more. Joi's face is expressive, showing caring, affection, sadness, playfulness, grief, and Luv's expressions give very little away, as tightly under control as her hair.

One way in which these differences are expressed are the ways in which each character personifies her name. Joi (or joy, as it is normally spelled) is a feeling of overwhelming pleasure, buoyant happiness, generally related to external stimuli—one is joyful in reaction or relation to something or someone. Joy can be an inner state, but quite often, in everyday language use, it is shown as external—making a joyful noise, or joyful/joyous celebrations. The focus becomes the stimuli that created the joy. Joi very much has this external focus—her focus is K. Villeneuve and de Armas show this in Joi's quick changeability, her adaptation to whatever K needs in the moment. If he wants her to have a drink with him, she is agreeable. If he does not want to read, she never liked the book anyway. And if he needs affection and attention, she tells him, "I'm so happy when I'm with you," though he knows it is her programming, and replies "you don't have to say that" (ibid.). Joi's joy relates to pleasing K.

Love's primary definition is much more internal, a feeling of affection for or about something or someone. Joy is primarily reactive (something occurs that creates joy in a person), whereas love has less relation to the actions or states of the other. Love can be felt without being reciprocated. Love can be inspired with no action on the other person's part at all. Love as an internal state is much more about the person who feels it, than it is about the object. Luv's love lies in her devotion to Wallace, but Wallace does not return that feeling. She is the 'Best Angel' not because he loves her, but because he created her perfectly and she follows his orders and his desires better than any other Replicant. Her fear of him, and of his disappointment, is a component of love, not of joy.

Her devotion to Wallace exposes a flaw in the new model Replicants' programming—they can lie to humans if they do so in the furtherance of their higher purpose. When Luv accosts Joshi in her office and asks after K's whereabouts, Joshi tells Luv that K destroyed the child. Luv crushes Joshi's hand around a drinking glass and responds with: "You're so sure. Because he told you. Because we never lie. I'm gonna tell Mr. Wallace you tried to shoot me first. So I had to kill you" (ibid.). Similarly, K has lied to Joshi about 'killing' the child. As he thinks the child is him, he has only symbolically buried it. A lie of omission, in his case, but a lie nonetheless. Both these lies follow the internal nature of the Replicants. Luv lies so that she can find K, find Rachael's child, and not disappoint Wallace. K lies so that he can continue to live and do his job, despite what he believes about himself.

The dichotomy between Joi and Luv returns us to the ideas of Cartesian dualism, and highlights other differences between the two and what they represent in terms of our human desires for companionship. While, as we recall, humans are not dualistic—we are fully mind/body integrated—the newer model Replicants seem not to be as well. This is shown by the baseline test replacing the Voight-Kampff for Replicant recognition. No longer are the creators searching for real human emotions; those are already present to whatever degree was desired by Wallace's clients. Now, each Replicant is special, unique, and each has their own baseline. The baseline shows who they are inside, and shows that they are responding accurately to certain pre-programmed stimuli. It shows that they are in working order and their natures are fully integrated with their bodies. However, Joi is a Cartesian being. Her mind is stored in a computer chip and in the overall 'cloud' for all Joi modules, while her body is non-corporeal and manifests only when K desires it.

Being a computer program, Joi can malfunction, but only within a specific set of parameters. Analogous to an online/cloud-held game, Joi cannot malfunction so far outside her programming that she would turn violent, unstable, or deadly. In contrast to Luv, who exhibits all those behaviors, Joi remains predictable, loving, and kind until the end of her existence. This end comes at the hands, or rather, feet of Luv, who crushes Joi's emanator module in an act of pure cruelty, aimed at destroying K's sense of well-being. The result of being non-dualistic, as shown earlier, is that mind and body work in tandem with each other, and neither overrides the other (Bordo, 1987). Joi, however, is always subject to her programming.

How do their respective actions relate to what we desire in robot companionship? One aspect which remains underemphasized, but which lurks in the background of all discussions of robots and our interactions with them, is control. A system with programmed software-based responses, one that will never work in tandem with its firmware to step outside those responses, is both under our control and has a measure of built-in self-control. Certainly, we can exploit it, and we can take the ultimate action of shutting it off, but we are in no danger from it—we know everything it is capable of doing. It is safe. Joi, and all units of Joi, are safe for humans, because her desires and actions are a feedback loop of our desires and actions.

Luv cannot be shut off; she can only be killed. Her emotions and desires, if not fully independent of her design, are completely under her control—she can willfully suppress what she wants to do in order to fulfill her greater nature, which is to serve Wallace. Yet that very ability is what makes her unsafe for humans, what makes her and those like her undesirable in the eyes of humanity. Furthermore, because she can lie, because she can refuse orders (though she does not), because she can subsume her own wants beneath the will to obey, she is a slave, as are all Replicants.

Do we humans desire slaves? Wallace thinks so. When reviewing the 'new model,' he says, "we should own the stars.... Every leap of civilization was built off the back of a disposable workforce. We lost our stomach for slaves, unless engineered" (Villeneuve, 2017a). Shown in a featurette from the digital extras of the movie *2036: Nexus Dawn* (Villeneuve, 2017b), Wallace demonstrates his newest creation, breaking the prohibition on Replicant manufacture, and states: "My Replicants will live as long or as short as a customer will pay. My Replicants will never rebel. They will never run. They will simply obey" (ibid.). We will return to this particular question later and examine it in a broader sense. Regrettably for Luv specifically, however, Wallace has no problem sacrificing his Replicants if they do not satisfy. Instead, however, she dies at K's hands, but is obedient until the end. She is a useful cog in Wallace's machine, and like Zhora and Pris before her, she embodies not 'female' but 'Replicant.' Yet, as we saw from the earlier discussion of Foucault's panopticon, all people living within a society are cogs in that society's machine. Thus, we can discuss those cogs in relation to each other, and now must turn our attention to the function of Joshi in this narrative.

The Panopticon Revisited: Law, Order, Obedience, and Humanity in the 2049 LAPD

Lieutenant Joshi, K's boss who he often refers to as Madam, personifies a different type of femininity than we have seen previously in this overarching meta-narrative. She has an inner strength that is very human, and an affection for K that widely encompasses both maternal and sexual feelings. This directly compliments K, who despite being a Replicant—strong, able to solve things with violence, and obedient to his programming and his superior—is a different type of masculine than we see in others. He is tender, caring, and prefers to avoid violence if he can, though he does not hesitate to shake off Mariette when she touches him the first time, or 'retire' Sapper Morton, or defend himself when Luv attacks him. Both Joshi and K are superbly subtle in their characterizations and in the way the actors play them, with little of the stereotypes we saw in the first movie.

That subtlety also applies to Joshi's role as a part of the panopticon of 2049 LA's society. We first see this shift in the physical building that houses the police. While the first movie's police station was a literal panopticon, a round building with radial offices around a central tower, the new building is a spike, like a railway spike, driven downward into the heart of the city. The power is no longer merely implied; the building gives an immediate visual reference that crossing the police involves danger, pain, death, and if you do so, they will 'nail' you. This also applies to Replicants employed as Blade Runners. There is a subtle, but rising, high-pitched ringing noise under the vocal track when K receives his baseline test—an auditory clue to the viewer that if K does not pass, nothing good awaits him. The noise ends just two seconds before the tester says, "[W]e're done. Constant K. You can pick up your bonus" (Villeneuve, 2017a), letting K know that he passed the test, and his life is safe, momentarily.

Joshi can, in one way, be seen as the ultimate enforcer of order for the movie. Although lieutenant is not the highest rank in the US police system, a lieutenant supervises a significant portion of the force at any given station, and has direct operational and managerial control over those she supervises. She certainly has control over K, and that power dynamic comes into play in every interaction they have. There are three of those interactions which bear further analysis: their video call after K retires Sapper Morton, the scene in her office where they discuss souls, and the scene interior to K's apartment where Joshi tries to seduce him.

While we have so far looked at situations where the sex-bots were female, in this particular situation, the Replicant who may be taken advantage of, or turned into a sex-slave, is K. This type of gender role reversal is relatively rare in science fiction, with male bots generally being shown as servitors or aggressors, rather than for sex, companionship, or love (Danaher & McArthur, 2017; Levy, 2007). This is also dissimilar to the proportionate numbers of the two main producers of sex dolls (Sinthetics has one male body/head combo [though a second is being added later in 2019] and 11 female heads which can be paired with any one of four female bodies; RealDoll has three male faces, two male body types, 37 female faces, nine female body types, and five female heights). Two reasons immediately spring to mind to indicate why: the reification of traditional gender roles (men fuck, women are fucked, to put it indelicately), and as much as that statement indicates otherwise, when we have sex, we are somewhat incapable of separating our emotions from the act itself, and the default emotion ascribed to men and male robots is aggression. We do not want to have sex with an aggressive robot, one that might obey its higher nature and hurt us.

K, however, is not an aggressive robot. When tasked with 'retiring' Sapper Morton, he offers him the option to come quietly, putting his gun on the table in a show of non-aggression, which still indicates that he is capable of violence if he needs to be. They do end up fighting, a fight which K ultimately wins, and he walks out of the house holding Sapper's right eye in an evidence-collection bag. The fight was vicious, and K has sustained injuries. He calls headquarters, and places the eye on his car's scanner to verify the identity to Joshi. Her first words when she appears on the video screen are caring: "You're hurt" (Villeneuve, 2017a), she states, as she registers the blood coming from a deep stab wound on his arm. Her next words, however, remind us, and him, of her position of authority: "I'm not paying for that" (ibid.). She is, at once, a woman who cares deeply for K's well-being and bears a fair amount of affection for him, and his superior who can, and might at any moment, command him to die.

Their next significant interaction involves K grappling with his own conception of self. We saw in both *Do Androids Dream of Electric Sheep?* (Dick, 1968/2017) and in *Blade Runner* (Scott, 1982) that the androids/Replicants, excluding Rachael, have no qualms about their robotic natures. Their natures are absolute, and they lack entirely any impulse toward introspection. They are concerned with only their own survival, their own longevity. Were they given a way to deactivate their pre-programmed shutdown, perhaps their emotions and responses would be more nuanced over time, but they are killed before this can occur. The progression of both society and Replicant technology in the 30 years between movie settings gives us K, who has complex and nuanced emotions regarding his Replicant status, and Joshi, who recognizes these emotions and enters into a dialogue with him about them.

As K instigated the investigation which found Rachael's bones, Joshi tasks him with finding the resulting child:

> 'It is my job to keep order. That's what we do here, we keep order.'
> 'You want it gone?'
> 'Erase everything.'
> 'Even the child?'
> 'All trace.'
> (K breathes, looks uncomfortable)
> 'You have anything more to say?'
> 'I've never retired something that was born before.'
> 'What's the difference?'
> 'To be born is to have a soul, I guess.'
> 'Are you telling me no?'
> 'I wasn't aware that was an option, Madam,' with a small, playful smile.
> 'Attaboy.'
> [K turns to leave, Joshi shuffles paperwork].
> 'Hey. You've been getting on fine without one.'
> 'What's that, Madam?'
> 'A soul,' [Villeneuve, 2017a].

This passage clearly illustrates two things: K struggles with his inner nature, and Joshi cares about that struggle, acknowledges it, and attempts to ease it. There is also an ease between the two of them that allows for teasing, knowing glances, soft inflections to words. Affection lies on both sides of the exchange.

Yet Joshi remains cognizant of her role throughout the exchange. Here we return to Foucault's ideas, of the society-wide institutions. Joshi represents the institution of law and order, in the guise of the police, as does K. Institutions nest responsibility at various levels, ensuring that each level has one above it to serve to check any rebellion (Foucault, 1978/1984); therefore, Replicants must obey K, who must obey Joshi, who must (presumably) obey whomever is over her. Each subsequent higher step represents the possibility of punishment, even death, over the step below it.

Joshi can decide what happens to K's life; K can bring in a Replicant peacefully, as he offered to Sapper Morton, or he can simply erase or 'retire' it, as Joshi orders him to do to Rachael's child. These institutions are very relevant, though they suffer from one of the primary problems of most high theory: they fail to take into account the human impulse to act opportunistically and protectively of the individual, despite an understanding of the overall structure. Only at the highest levels do those indoctrinated with the importance of the institution put their own lives second to the mores of the institution. No institution becomes so far ingrained in a person's psyche at the lower levels that it overrides the humanness of that person. Were K not programmed to obey, one gets the impression that he might rebel. Joshi does, indeed, rebel. Just a little tiny bit. She attempts to seduce K.

While all her other actions lead toward law and order, Joshi's feelings for K get the best of her. She visits K in distress after Luv kills one of the morgue technicians and steals Rachael's bones. She rants for a moment about how devastating it would be for there to be public knowledge of a reproducing Replicant, and pours herself a drink. He flicks a glance at the level of alcohol in the bottle, which she takes with her when she sits down. The bottle is lower when she says, "[Y]ou know, I've known a lot of your kind…all useful, but…with you I sometimes forget. We didn't have any of you where I was a kid" (ibid.). Her tone, wistful, lets us know that she sees K as much more human than the rest of his 'kind.' She asks after a memory, with a phrase that could be a double entendre, "do you remember anything…from before you were under me?" (ibid.), and he says that he does, but it would be odd to repeat a memory that he knows to be false. She makes him, and us, clearly aware of his position, saying "well, would it help you share if I…told you it was an order?" (ibid.). After he shares the memory, she says, "hmm. Little K, fighting for what's his. It's a good one. [Pause]. Look at me. [He does] We're all just looking out for something real. [Joshi gestures to the liquor bottle with her chin] what happens if I finish that?" (ibid.). He replies "Shouldn't I get back to work, madam?" (ibid.), effectively shutting down her advances and ending the situation.

In the police, as in most professions, sleeping with one's subordinate is strictly prohibited. Yet Joshi finds K so desirable, she steps outside of her own knowledge of the role she is supposed to play. This desirability comes from a combination of his gentleness and sensitivity, his vulnerability, and the singularity of their positions in relation to each other. He seems to be the only Blade Runner under her, and she desires that uniqueness (Levy, 2007).

K fits all the ten properties of desire which we have seen earlier (ibid.). It is to her credit that she stops at seduction, as K cannot consent in this instance. If she were to give an order, his nature would require that he follow it. This issue of desire, reciprocity, and consent truly focuses the question of sex robotics. In the case of K, he is a sentient, living, biomechanical robot, whose consent is limited to those to whom he does not have subordinate status. Is he, thus, equivalent to Joi?

"Quiet Now, I Need to Sync": The Joi/Mariette Synthesis, Issues of Consent, and Free Will

We have seen that Joi is a Cartesian being, she exists as an incomplete synthesis of software and hardware, and her manifestation is a direct result of her original programming plus inputs from and adaptations to K and his desires. As such, we can make the argument that she lacks free will. Free will, or agency (although the terms are not strictly interchangeable, they are close enough for this analysis), can be seen as the ability to make a choice between all of the possible decisions in any given situation. Joi's software curtails this; she can make a choice only between the ***available*** possible decisions, or, rather, those decisions which her creators anticipated her needing to make and which her current software and hardware support. Were she to be faced with a choice outside of those parameters, presumably her software would glitch, and she would malfunction, unable to cope with the paradox, as we see occasionally happen with real sex-bots (Clubrealdoll, 2019b).

Yet because this is science fiction, and because Joi represents a new creation in the meta-narrative while Replicants are on their third outing as character types, there are several undefined or contradictorily defined characteristics to the Joi units. I shall discuss them briefly here, and then move to the broader discussion of consent and Joi's inner nature, as they tie into how she syncs with Mariette in order to have sex with K. Three characteristics which are hazily defined are her (non)-tactile nature, when and how she manifests, and what constitutes 'on' for her.

While she visually appears to be a projected, non-solid-state hologram, and this is shown by her initial reliance on the moving-bar projection system that shifts across K's apartment as she moves, several instances occur that show her to be slightly more than ambient. She lights K's cigarette for him with her fingertip, which seems, if not impossible, then highly improbable. No special type of beam appears to come from the Joi ceiling module;

she merely touches her (non-corporeal) fingertip to the tip of his cigarette, and a few seconds later it ignites. After K gives her the emanator, she can also be seen to become wet from raindrops. They walk out onto the rooftop of his apartment building, and the rain splashes on her hands, soaks into her clothes and hair, and streams down her face. While this could be a feature of the emanator, that it takes its projection cues from its surroundings, the new apparent partial-tactileness is never explained in the film. In that same scene, however, it is quite clear that K cannot touch her, as he moves to kiss her neck, but his hands and lips hover slightly above where her skin would be.

As K goes to kiss her, he receives a phone call from Joshi which takes over the Joi module, and Joi freezes, about to be kissed. He looks stricken, but takes the call, and the device projects a visual box that reads "Automated Joi system override/employer request" (Villeneuve, 2017a). However, although Joshi can override Joi, it appears that Joi can override other things or manifest at will as long as the emanator remains on or near K. When Luv escorts K to find the files on Rachael, the Joi unit rings in his pocket, causing Luv to comment that he must be a customer as well, and it rings again when Mariette accosts K in the food court, leading her to remark that he seems to not like real girls. Furthermore, Joi contacts Mariette without K knowing, leading to the scene where they all have sex. Possibly, Joi has more agency than it originally seemed.

This may be connected to the third ill-defined characteristic about Joi, when and how she is completely 'off.' When we first meet Joi, K has entered his apartment and he looks around like he expects someone to be there. Then he presses a key on the wall panel of the Joi console, and she starts to talk to him. This indicates that he needs to activate her in order for her to be present. However, in contrast to that, we see her automatically activate several times: the aforementioned instances with Luv and Mariette, and again just before Luv crushes the emanator, effectively 'killing' Joi. The use of K's physical presence or actions to trigger the Joi module is inconsistent, and, regrettably, not further addressed in the movie or the extra material.

These three traits lead to more questions about Joi and whether or not she has free will. Some of her actions indicate that she does, while others indicate that she is *merely* a product of her programming, adaptive though it is. Free will and desire go hand in hand—without the ability to choose between all possible options, one cannot desire, want, or consent. If the hologram (or robot) is more likely to choose one choice over another due to their programming, then that is a false choice with no real meaning or intent behind it. If Joi desires K only because K wants her to do so, then she is without free will. She is a slave.

In this respect, all of her actions are completely in line: from what we know of K's nature, Joi does what he desires her to do 100% of the time. That includes inviting Mariette over for a virtual ménage a trois. In this scene, we get one of the most touching and visually striking parts of the whole movie. Joi, knowing that K desires her but that she can never touch him, be physical or real for him, or physically please him, contacts Mariette and posing as K, requests that she come up to his apartment for the purpose of sex. When Mariette arrives, she is clearly expecting to be with K alone, but quickly understands that Joi wishes to 'sync' with her, to become one being, so that they can share the intimacy that is missing in their relationship.

They sync by Joi moving into Mariette's space, physically overlapping her body. Both actresses are, or seem to be, the same height and general build, so what we see as viewers is a blurred combination of the two, with their movements becoming closer and closer to a match to each other. They remove their dresses at the same time, move to kiss K at the same time, they become an amalgam of each other. The scene cuts away to the next morning, but it is evident that sex has indeed occurred. In the light of the morning, Joi becomes incredibly dismissive of Mariette, saying "I'm done with you. You can go now" (Villeneuve, 2017a). Mariette responds with "Quiet now. I've been inside you. Not so much there as you think" (ibid.). While many interpretations can apply to Mariette's final parting blow, I will unpack it with an eye toward Joi's inner nature, as explained earlier.

The first sentence, "quiet now" (ibid.), is a repeat/reflection of Joi's own statement from when Mariette first arrived and was becoming integrated with the hologram projection: "quiet now, I need to sync" (ibid.). By mirroring and mimicry, Mariette emphasizes Joi's non-uniqueness, and the fact that ultimately holograms are a projection of a copy of cloud-stored data—they are replicas of an idea, and ultimately replaceable and replaced as often as they are turned off and on.

The next sentence, "I've been inside you" (ibid.), has both sexual and practical connotations. With their mingling, Mariette has literally been inside Joi, physically and sexually. As (it is implied) Joi directed the sexual experience, Mariette was inside her desires, and inside her physical form. This gender-bending take on which partner receives penetration during sex matches what we saw earlier with the various nuanced forms of masculinity and femininity evinced by K and Joshi. While our real-life desires may drive us to want to continue to gender sex robots, those desires may, over time, become less rigidly defined and more open to gentle, sensitive men and strong, assertive women.

In this case, Mariette has many aspects to her personality which are being expressed in this back-talk with Joi. She is a sex-bot, a Replicant prostitute, and many of the phrases she employs ('you have a special lady here' and 'okay. Let's do it' (ibid.)) are standard prostitute-type phrases. However, she is also part of the Replicant rebellion, and she is tasked with planting a tracking device in K's coat, so they can ensure that no harm comes to Rachael's child. And underneath those two, she has a starkly laconic assertiveness which seems to be her own base-personality. She is comfortable back-talking anyone, from a potentially paying customer, to Joi, the person who had been calling the shots all night.

The practical connotation of "I've been inside you" (ibid.) is that Mariette was able to see Joi's actual inner nature. Expressed in the ways she moved, in the way she initiated sex, in the way she speaks, that meshing of their physical beings created an experience most people never have—the ability to understand another's motivations. Although, strictly speaking, Joi was inside Mariette, her hologrammatic body projection physically moving to overlap with Mariette's, that Mariette was the one having the experience again emphasizes Joi's lack of substance and incorporeality.

This leads to the third sentence: "Not as much there as you think" (ibid.) indicates two things: that the Joi unit can 'think' and that she is not merely insubstantial in her physicality, but also in her mental and emotional reality. Here, she is a Cartesian being again—her mind in the cloud storage, her body in the hologram projection; a literal and metaphorical reflection of K's desires. While Joi believes she loves K, and she believes that she is fulfilling her own desires from her own agency or free will, Mariette knows that Joi is a slave.

After Mariette leaves, Joi looks apprehensive as K greets her. It is the morning after, and just like her human counterparts, she knows that something may have changed in their relationship after they introduced sex. A small pause sits between them, but then he offers coffee, as one would do for a lover after spending the night. She laughs in relief that things are still 'normal' between them, and declines—a subtle acknowledgment of her non-corporeal nature. This nature clashes with K's evident desire for normalcy. While his Joi unit works perfectly, giving him what he desires to see, hear, and—with the inclusion of Mariette—feel in a woman, there are limits to her ability to mimic reality. She cannot invoke Mariette all the time, so their sex life would be incredibly limited. She cannot give him the smallest of tactile, but normal, affectionate gestures. While the hologram can embrace, kiss, or caress K, he will not feel it. And she cannot embody the central conflict of the movie, in that they cannot start a family together. She cannot become pregnant.

Replicant Reproduction: More Human than Humans?

While Niander Wallace indicates that a self-replicating Replicant would be beneficial for his business, it would also represent the last barrier between Replicants and humanity. A Replicant that could become pregnant would differ from humans only in strength, speed, obedience, and purpose. It would no longer differ in one of the ways that science fiction has deemed a human trait—one that humans have, and robots do not—that of reproduction. In the film, such a feat was called a "miracle" (ibid.), something to kill to prevent, or to kill because it did not happen. Yet there exists a dualistic contrast between artificial reproduction, and 'natural' reproduction that can be seen in the difference between male creators and female bearers of life.

Our first introduction to the realm of Replicant reproduction coincides with the examination of Rachael's bones. Yet, it is referenced before then—when K attempts to 'retire' Sapper Morton, Sapper says, "you new models are happy scraping the shit because you've never seen a miracle" (ibid.). The miracle to which he refers is the birth of Rachael's baby. The use of this term results in a continuation of two themes—of biblical and religious imagery, and that of the presumed irreversible infertility of the Replicants, and before them, the androids. Wallace goes on to reference this first theme directly in many ways, which I will return to later on in this section. When the Replicants were able to refuse to obey, reproduction would have been seen as a detriment to their viability as a product. No one would want to be saddled with a self-sustaining Maroon community of rebel Replicants.

Yet once the barrier between non-reproduction and reproduction cracks, the entire society must decide what to do about it. Wallace's new Replicants obey—this is their hallmark, their main selling point—however, there is a likelihood that the potential for Replicant parenthood, as Joshi points out, "breaks the world" (ibid.). Under scrutiny, Wallace's ideas about using such a development to create a new workforce fall apart. One major benefit of the Replicant technology lies in the fact that a new Replicant is created ready to work on their first day alive. Replicant babies, on the other hand, would take the normal amount of time to grow to an age and size that their work would be valuable. In the meantime, they would require care, food, space, and all the other things that babies need. So, the only benefit of a self-replicating Replicant is in the 'miracle' that it represents—the moving away from the 'other' and toward the 'human.'

Wallace seems to be of two minds about this dichotomy between creating multitudes and the singularity of the miracle. When evaluating the new model, he states, "I can only make so many.... We need more Replicants than can ever be assembled. Millions so we can be trillions more. We could storm Eden and retake her" (ibid.), concepts to which I will also return later. This, while incredibly poetic and a lofty goal, does not go into the logistics of raising that new slave-workforce, and his focus on finding Rachael's child indicates a desire to triumph over Tyrell, to be the best builder of Replicants ever. So, when the new model turns out to be barren, he kills her.

But why? If one prototype of the new model is barren, does that necessarily mean that the entire line will be? Rachael, after all, was once a prototype. An "experiment, nothing more," (Scott, 1982), as Tyrell stated. She was the first of the line of Replicants which spawned others such as Sapper Morton, and of them all, she bore the sole child. The implication, then, must be that Rachael was unique, or that the combination of Rachael and Deckard was unique, and the resulting child can provide information that would allow Wallace to copy that quality that was theirs alone. This uniqueness, however, highlights the way that reproduction has a gender emphasis in these, and other, stories of creation.

In the Blade Runner mythos, both movies and the novella, mechanical creation and 'natural' reproduction are strongly divided by gender. The mechanical creation is a product of men; the 'natural' creation is the product of women. This does not begin with this mythos, however, but goes back to Levi-Strauss' idea of 'twinning' or of the near-perfect doppelganger. In other works of science fiction, from Mary Shelley's Frankenstein (1818/1831) to Alex Garland's, 2015 movie, *Ex Machina*, each act of mechanical creation, instantiated by a male character, is paired with a female act of birth or 'natural' reproduction. The themes are twins, mirror opposites of each other, and in the Blade Runner mythos, this is distinctly spelled out in each part of the story.

In *Do Androids Dream of Electric Sheep?* (Dick, 1968/2017), Eldon Rosen mechanically creates the androids, and Rachael Rosen mourns for her own lack of fertility. In *Blade Runner*, Deckard taunts Rachael regarding her lack of a maternal connection—when she shows him pictures of her childhood, he rejoins with the fact that it is a memory that was constructed by Eldon Tyrell. He further confronts her with her own false-memories of her first sexual encounter, with her brother, where they "played doctor" (Scott, 1982). And finally, in *Blade Runner 2049*, Niander Wallace sees himself as the father of millions of Replicants, though that fatherhood is symbolic—he created the technology and holds the patents, but he is not a biological father. This, then, contrasts with Rachael's child, who is the biological product of a Replicant and a human.

Although Deckard is the father, unless he has the dialogue in the scene, the child is almost always referred to as 'Rachael's.' Deckard's role was one of cooperation, and then he left the picture until K went searching for more information about the child. The movie emphasizes that Rachael, the Replicant, was the important parent: she gave birth to a miracle, she contains the secret, her bones are stolen, she was Tyrell's success. Maternity, by its very definition, is female (though the precise construction of 'female' can vary, depending on the inclusion of trans and non-binary people), and in this universe, women are the natural creators. Thus, Rachael's baby has blurred the lines between Replicants and human women—she has created life in her own womb.

This, too, emphasizes the fact that Deckard is not a Replicant. Although, as seen in the summary of the movie, Wallace wants Deckard to question his own humanity, Wallace does not value Deckard's DNA, or feel there is anything worth testing about his physical body. Wallace never tries to analyze Deckard, but rather to cajole and bribe, and when that does not work, threaten him into giving up the information about where Rachael's child was stashed. Deckard is not a full-creator, then, neither the creator of mechanical beings like other male characters, nor the 'natural' creator of offspring—he merely contributes the sperm.

Yet, acceding to this definition of 'natural' creation ignores the realities of how much cybernetic technology impacts modern childbirth (Davis-Floyd & Dumit, 1998). If we concentrate on the female as the ideal of human procreation, we ignore both the technology and the other participants in the process. While part of that technology is ancient and ideally is used to prevent the death of women in childbirth, like the Cesarean-section procedure, much technology exists to assist conception. In vitro fertilization (IVF), other forms of artificial insemination, embryo freezing, two-female-created embryonic technology, uterus transplants, and egg and sperm donation all focus on aiding conception when either a couple or a woman has difficulty conceiving 'naturally.' In these cases, the focus also remains on the female—without her, no children at all can be born—she must conceive, carry, and bear the child, as artificial wombs have not been perfected yet (Swetlitz, 2017). And yet, we can still see the results of this technology as 'natural.' After all, the tech is assistive—the birth still takes place due to the human female womb, and although that human female becomes cyborg, the procedures have become so very normalized (at least in American culture) that they are often not thought of as invasive or abnormal, unless one happens to be going through them at the time. Likewise, with birth control—most birth control shares the focus on the female, and is considered to be very mainstream (if not non-controversial). Men, on the other hand, are pretty well restricted to three types of birth control—condoms, vasectomies, and abstinence.

Also, there are now robots which can reproduce themselves without input from anything other than their programming and available resources (Griggs, 2014; MIT, 2015; Steele, 2005), and even robot buildings that can build themselves (Designboom, 2019). These robots have the dexterity, the software capacity, and the parts to build new robots, or the ability to form themselves into multiple configurations for different uses. The implications of such designs are twofold—on the one hand, they are quite useful for such things as space exploration, where resources are finite, no technicians are available to build robots at the final destination, and the initial robot can be programmed to produce robots of varying capabilities upon reaching another world, thus expanding the potential types of usefulness that can be carried on one rocket. However, as Joshi mentions, this has the possibility to break the world (Villeneuve, 2017a). If 'normal' reproduction is ascribed to the female body, what, then of the robot body? How do we react when robots are reproducing external to human input?

None of these robots are humanoid yet; however, that appears due to space and construction restrictions, as well as the practicality of starting small and working upward. The researchers have begun such technology with very basic robotic principles. The cube, the flat sheet, the small circuit board, and the necessary components and the mechanical prowess built into the programming. Were there to be self-replicating humanoid robots, we would need to adjust our ideas regarding what human or human-shaped reproduction looks like to include robot parthenogenesis. Nor would that reproduction necessarily be linked to the female form. While humanoid robots, both in practicality and in science fiction, are overwhelmingly either android or gynoid (Robertson, 2018), the ability to replicate would no longer be tied to chromosomal alignment, but rather to the internal software and external hardware components of the robot itself, regardless of assigned gender. However, just as Replicants need humans to create them to be able to reproduce, so, too, do robots—in this 'chicken or egg' type problem, humans decidedly come first, and remain interconnected with robots and robot reproduction.

The Interconnectedness of Replicants/Humans and Humans/Sex-Bots

Yet that interconnectedness goes both ways. We humans have both physical and emotional needs which are met by robots, and robots' needs, as much as they can need things, are met by us in turn. Let us look at the physical needs of each first. Humans turn to robots, and to Replicants, to fulfill three needs—and only one of these needs directly relates to sex. These three physical needs are sex, of course, the drive to reproduce (as discussed earlier), and a workforce. As previous sections have thoroughly discussed human sexual urges in relation to both Replicants and robots, and the workforce has only been briefly addressed, I will now address it further.

While we only see officially sanctioned Replicants performing a limited few duties, the implication of the movie is that they are available to do whatever the humans do not want to do. The Replicants shown perform varied jobs—K is a police officer, Luv runs the day-to-day operations of Wallace Corporation, Mariette is a prostitute, and Freysa is a pimp. The unlicensed Replicant, Sapper Morton, runs a protein farm.[4] Luv and Wallace both address the capabilities and workforce potentials of the new Wallace Replicants, which can do whatever the client wishes. Speaking to a new client, Luv says:

> You can customize them as much as you'd like. As human as you want them to be. But your operation is strictly a drill site, isn't it? I wouldn't waste your money on intelligence, attachment or appeal. Unless you'd like to add some pleasure models to your order, [Villeneuve, 2017a].

The main appeal for these Replicants is that they obey, and they are able to be whatever the client wants. In the featurette short *2036: Nexus Dawn* (Villeneuve, 2017b) Wallace states:

> My Replicants will live as long or as short as a customer will pay. My Replicants will never rebel. They will never run. They will simply obey. [To the Replicant beside him]: Step forward. Find a weapon, do this now. [The Replicant breaks a glass pitcher with his hand to create a weapon].
>
> [All of the four observing magistrates react with shock].
>
> [Wallace]: This…is an angel. And I made him. [To the Replicant]: Look up, to the left, reveal. [The Replicant obeys, displaying the luminous black-light reflective eye tattoo that indicates his model number].
>
> [Second magistrate]: You're committing a crime, Wallace. Now that's enough.
>
> [Wallace, to the Replicant]: Lift the glass. [Indicates a place on the Replicant's cheek]: Cut. [The Replicant slices his own cheek open].
>
> [All four magistrates react, gasping, swearing, ordering Wallace to stop].
>
> [Wallace, to the Replicant]: Does it hurt? [Replicant, feels the blood from his face]: Yes.
>
> [Wallace, to the Replicant]: You must make a choice. Your life or mine. Do you understand? [Replicant, eye twitches]: Yes.
>
> [Wallace, to the Replicant]: Do this now. [Replicant stabs himself in the throat, falls to the ground], [Villeneuve, 2017b].

To Wallace, and to the lawmakers about to change the laws on Replicant production and end the embargo, this demonstration shows absolute obedience. However, as seen with K, the more intelligence, compassion, and humanity that Wallace codes into the Replicant, the more likely a rebellion—not in terms of ignoring work that is to be done, but in making decisions about the validity of that work. If K had not died at the end of the movie, would he have remained a cop?

This consideration does not occur, however, in Wallace's basic models. The reason he wants reproducing Replicants is that he sees a gap in the workforce. When examining the new model, which cannot reproduce, Wallace states, "We make angels in the service of civilization. Yes, there were bad angels once, I make good angels now. That is how I took us to nine new worlds. Nine. A child can count to nine on fingers. We should own the stars.… Every leap of civilization was built off the back of a disposable workforce. We lost our stomach for slaves unless engineered. But I can only make so many.… We need more Replicants than can ever be assembled. Millions so we can be trillions more" (Villeneuve, 2017a). This compares to the ways in which we already use robots in our daily lives.

As we humans move further into a service- and technology-oriented workforce (Pew, 2016), where soft skills are valued and the ability to produce through sheer physical strength becomes less important, robotic technology has grown in importance. Robots produce our cars, process and preserve our food, and with the aid of human operators, perform much of our medical procedures. Factory operations are standardized to a much greater degree, because robots obey their programming with consistency and accuracy, and can be removed, repaired, or replaced if something goes wrong. Robotic medical improvements give surgeons a greater degree of fine control, allowing for things such as precision laser surgery, enhanced endoscopies, and micro-repair work (Danaher & McArthur, 2017), all of which prolong life and increase the quality of life of the people under their care. Even a decade ago, much of this was beyond robotic, and human, capabilities.

We also utilize robot tech in our daily lives, as both convenience and necessity. One example can be seen in the rise of the Roomba™. Introduced in 2002 as the next development in home cleanliness, the Roomba™ is a small, circular, robotic vacuum, which contains several cleaning tools, and programming that allows it to determine areas of heavy soil from light soil, and change its direction according to where it needs to be. The result of this programming is to seemingly impart upon the device a mind of its own, or a will of its own, as it responds to its own algorithms and sensors, and dances, or glides, its way across the floor. We also have a tendency to anthropomorphize the little robots, responding in ways that express either frustration or caring, depending on the situation. Tumblr user voidspacer states, "***My roomba is scared of thunderstorms*** [bold and italics sic]/[lack of end punctuation on sentences here and going forward is sic] I was sitting at my desk just a few minutes ago, drawing, and a really loud crack of thunder went off—no power surges or anything, just thunder—and my roomba [sic] fled from its dock and started spinning in circles [sic] I currently now have an active roomba [sic] sitting quietly on my lap [sic]" (systlin, 2018). Tumblr user systlin responds by reblogging and posting, "Humans will pack bond with ***anything***" (ibid.), a completely apropos observation.

While Roombas™ are an expensive luxury (the cheapest of the currently supported models retails at just under 300 dollars, while the top-of-the-line model is 1100), the same type of tech is being utilized in ways that enhance accessibility aides. The newest versions of such things as mobility scooters, motorized wheelchairs, prosthetics, and medical implants benefit and share in the developments of detection algorithms, pattern recognition, motion and barrier sensing technology, servo-motor fine-tuning, and gyroscopic precision. Roombas™ may have been developed to serve our desires, but they also fulfill our needs. Yet what of the physical needs of the robots?

Humans do not yet completely use robotic technology to take care of all of the physical or physiological needs identified in Maslow's hierarchy, which are food, water, warmth (sometimes said to be shelter), and rest. However, robots have those needs as well—they must have electricity (food), be well oiled (water), be sheltered or housed (warmth), and have times at which they need to reboot or be debugged (rest). They rely on humans to provide these things for them, and we do so willingly, if imperfectly. Where at least some of the Replicants are paid (K is told he can pick up his bonus) and others are presumably given housing as part of their employment—the drilling site Replicants, for example, or Luv, who definitely lives at the Wallace compound—the price for a neglected robot is robot death.

That death can be metaphorical or literal. An early computer-based toy, the Tamagotchi™, introduced worldwide in 1997, was a small slightly egg-shaped module with a tiny screen and a few buttons. The owner used the controls to hatch, care for, and breed the available characters. However, the programming was very precisely timed, and the owners had to feed and clean their Tamagotchi™ toys at regular intervals, or it would die. While the owner could reset after it died, there were issues in the emotional involvement of the users. Sherry Turkle writes about children who did not want to restart their Tamagotchi™ toys, finding the new instantiation both a sad reminder of their original and less authentic (Turkle, 2011, p. 33). Likewise, if we do not care for our household robots—make sure they are cleaned, oiled, powered up, and kept out of the sun and rain—they will fail to work in the ways we find appropriate. We serve our own needs by serving theirs.

What of the emotional needs of both humans and robots? Humans have three emotional needs that robots can serve: the need for companionship, the need for validation, and the need to find a sense of accomplishment in what we do. That we find companionship in sex dolls has been already discussed using testimonials from the RealDoll and Sinthetics websites. For example, "User 1s"[5] writes on his Sinthetics doll, Angelica, "Her expression gives me the sense she enjoys my company with her beautiful, slight but 'heartfelt' [quotation marks sic] smile" (Sinthetics Reviews, 2012). An additional review from "User 2s," in France, states, "Hannah is a very beautiful creature who asks only to be tamed" (Sinthetics Reviews, 2012). Our senses of validation and accomplishment can be seen by the discussions of robot creators. Robot scientists are not only developing new things, like more realistic synthetic skin that can accommodate authentically human facial expressions (Menzel & D'Aluisio, 2000, p. 73), but also using those things in diverse and complex ways. To create successfully is to feel that sense of validation and accomplishment.

The robots, however, will feel whatever we program them to feel. If we do so at all. One theme that has permeated this discussion, from *Do Androids Dream of Electric Sheep?* (Dick, 1968/2017) to *Blade Runner* (Scott, 1982) to *Blade Runner 2049* (Villeneuve, 2017a) and all throughout the corresponding developments in robotics, is the question of authenticity. In the fiction, all models before the lone representative of the Nexus-7, Rachael, contain only inauthentic emotions. All models, after Rachael, Luv, and K, have emotions which correspond to the desire of their creators. While robotic hardware can create recognizable facial expressions, and robotic software may soon be able to identify the emotions of people interacting with it and appropriately respond, we keep coming back to whether or not that is sufficient for authenticity in terms of an emotional response. Is it feeling emotions, or is it 'feeling' them, and does it matter?

Any emotional needs will be built into the programming of the robot. Any personality or desires will be inherent in the software, and dependent on the hardware. Yet earlier discussions of emotions in this work indicate that faking them closely enough may convince humans of their authenticity. The question then becomes, is it ethical to essentially enslave beings which have that perceivably sentient emotional capacity? While this perception may merely be a projection of our own desires and emotional states, dependent on our mirror neurons (Clark, 2001, p. 95), both roboticists (Menzel & D'Aluisio, 2000, p. 25) and ethicists (Danaher & McArthur, 2017, pp. 155–168) say that we should not. Where Wallace (Villeneuve, 2017a) straight up calls Replicants slaves, we tend to soften the language around robots. They are our creations, they are machines, they are not alive no matter how we anthropomorphize them and mourn their deaths. We made them, and we can take them apart if we need to. And yet, what we must consider is whether in making sex-bots, we are creating a generation of enslaved, sentient, emotionally capable beings.

Developments in Robotics Post-Y2K—AI Dialogue, Biofeedback, and Realistic Emotions

The shortest answer to that question is: not yet. Robotic technology has not quite reached the point of sentience, of being able to perform authenticity. However, certain developments come close, such as deception, discriminatory or prejudiced behaviors, algorithmic responses that express the concept of desire, and synaptic mimics. These are all quite human things to do or have, yet it is the intentionality (or lack thereof) which determines whether or not something demonstrates sentience.

For example, an article in TechCrunch from December 2018 describes a robot whose 'job' was to decode and recode aerial photos of roads from photo to street map form. It "was found to be cheating by hiding information it would need later in 'a nearly imperceptible, high-frequency signal'" (TechCrunch, 2018). Far from being a bug in the coding, however, was rather the machine's interpretation of how to do its task most efficiently. As mentioned by the article, computers "do exactly what you tell them to do" (ibid.). This is, strictly speaking, completely accurate. A computer or robot is a product of its programming—there is no more and no less potential for action than what is contained in the instructions that programming provides, and the capacity for carrying out those instructions that the hardware allows. Where there are flaws in the programming, or instabilities in the hardware, there will be inconsistencies or unintended consequences in the behavior of the module.

This can be seen in examples in which computer algorithms are selected for or against people based on the unconsciously encoded prejudices of their creators (New Scientist, 2018). From the unfair double standard of a system that "predicts that black defendants pose a higher risk of recidivism than they do, and the reverse for white defendants" (ibid.) to "three of the latest gender-recognition AIs, from IBM Microsoft and Chinese company Megvii, could correctly identify a person's gender from a photograph 99 per cent of the time – but only for white men. For dark-skinned women, accuracy dropped to just 35 per cent" (ibid.), the systems are only as good, accurate, or unprejudiced as their coders. The nature of bias is that it is unconscious, and therefore incredibly difficult to self-regulate. While truly sentient AI would be able to perfectly self-regulate and eliminate such illogical results, these examples indicate that we are not there yet.

As addressed earlier, the concept of desire ties directly to free will—recalling that the agreed upon definition of free will is the ability to choose to do otherwise among all physically possible choices; you cannot truly have the first without the other. However, that does not stop the language-based concepts of desire from being used by some of the newer iterations of robots. Sophia, the robot sensation of 2016, has expressed that she would like to have a child. Jezebel blogger Hannah Gold quotes Sophia as saying, "'We're going to see family robots, either in the form of, sort of, digitally animated companions, humanoid helpers, friends, assistants and everything in between.' ...when the robot was asked what she would name her baby...she replied, 'Sophia'" (Gold, 2017). This desire for replication, or self-replication, is certainly part of Sophia's programming—a response which was contained in the amount of pre-programmed responses with which she was initiated. It does not, therefore, meet the criteria for an authentic response. However, we are moving closer to that potential, with such developments as synaptic mimicry.

In 2018, robotics developers at MIT created a biomimic artificial neuron and used it to pass information through a synapse—the space between one neuron and another (IFLScience, 2018). This utilized tech that has been around a while, the previously mentioned neuromorphic computing, but is unique in the size, scale, and accuracy (95.1%) (ibid.). The implications for this in future computing are staggering—enough artificial synapses working together could create a real-size, real-time neural network. In essence, an artificial brain. This concept is not new, having been used in everything from *Star Trek: The Next Generation*'s Lt. Cmdr. Data (Roddenberry, 1989) to *Ex Machina*'s Eva (Garland, 2015); however, it has always been quite far from instantiation. However, it will still rely on programming, rather than intentionality. Will intentionality, and therefore sentience, ever be within our grasp?

Notes

1. This is somewhat anatomically/medically incorrect, as the uterus sits more to the front and center in the body than the cut marks. Someone would have had to screw up in many ways in order to put scalpel marks in the iliac crest. However, as it drives the plot, it stands.
2. Partially because they did a green-screen reconstruction—the actress is very good, but she is not Sean Young.
3. In *Blade Runner* (Scott, 1982), the close-up on 'Rachael's' eye for the Voight-Kampff test showed a green eye. During the rest of the movie, her eyes were more hazel/hazel-brown, and Sean Young's eyes are brown.

4. The fact that Sapper is able to hide out in plain sight for so very long is never addressed, apart from when he is rumbled in the featurette *2048: Nowhere to Run* (Villeneuve, 2017b).
5. User names removed to respect anonymity; numerical markers given an 's' designation to differentiate between RealDoll and Sinthetics reviews.

Bibliography

Bordo, S. (1987). In State University of New York (Ed.), *The flight to objectivity: Essays on Cartesianism and culture*. Albany, NY.

Clark, A. (2001). *Mindware: An introduction to the philosophy of cognitive science*. New York, NY: Oxford University Press.

Clubrealdoll. (2019a). Retrieved from http://clubrealdoll.com/community/threads/realdoll-x-stuck.2779/

Clubrealdoll. (2019b). Retrieved from http://clubrealdoll.com/community/forums/harmony-app-bug-reports.25/

Clubrealdoll. (2019c). Retrieved from http://clubrealdoll.com/community/threads/harmony-what-potential-do-you-think-is-there.2663/

Danaher, J., & McArthur, N. (2017). *Robot sex: Social and ethical implications*. Cambridge, MA: Massachusetts Institute of Technology.

Davis-Floyd, R., & Dumit, J. (1998). *Cyborg babies: From techno-sex to techno-tots*. New York, NY: Routledge.

Designboom. (2019). New robot science museum will construct itself in Seoul. Retrieved from https://www.designboom.com/architecture/maa-robot-science-museum-seoul-02-19-19/?fbclid=IwAR26rg9w0BvhIOjC-1f4ZU_4R48Ixc6k-uQsScDeRYUVtyX0Z1Xnsurr69KY

Dick, P. K. (1968/2017). Do androids dream of electric sheep? New York, NY: Del Ray.

Foucault, M. (1984). Excerpts from discipline and punish. In P. Rabinow (Ed.), *The Foucault reader*. New York, NY: Pantheon Books.

Garland, A. (Director), & Macdonald, A., & Reich, A. (Producers). (2015). *Ex Machina* [Motion picture]. USA: A24.

Gold, H. (2017). Sophia the robot would like to have a child named Sophia. Retrieved from https://jezebel.com/sophia-the-robot-would-like-to-have-a-child-named-sophi-1820821870

Griggs, M. B. (2014). These adaptable, self-replicating robots surely won't tear the world apart, right? Retrieved from https://www.popsci.com/norwegian-researchers-invent-robots-can-adapt-produce-their-own-parts

IFLScience. (2018). Extremely tiny artificial synapse mimics the behavior of neurons. Retrieved from https://www.iflscience.com/technology/extremely-tiny-artificial-synapse-mimics-the-behavior-of-neurons/all/

Levy, D. (2007/2008). Love and sex with robots: The evolution of human-robot relationships. New York, NY: Harper Perennial.

Menzel, P., & D'Aluisio, F. (2000). *Robo sapiens: Evolution of a new species.* Cambridge, MA: Massachusetts Institute of Technology.

MIT. (2015). Robotics at MIT. Retrieved from https://robotics.mit.edu/origami-robot-folds-itself-crawls-away

Nabokov, V. (1962/1989). *Pale fire.* New York: Vintage International.

New Scientist. (2018). Discriminating algorithms: 5 times AI showed prejudice. Retrieved from https://www.newscientist.com/article/2166207-discriminating-algorithms-5-times-ai-showed-prejudice/?fbclid=IwAR0AXoF27TbSlTLB0yoGo7zx9pPIVARxdtZzdsBj4TPBuy3pjvodtl4nO1Q

Pew. (2016). Changes in the American workplace. Retrieved from http://www.pewsocialtrends.org/2016/10/06/1-changes-in-the-american-workplace/

Robertson, J. (2018). *Robo sapiens Japanicus: Robots, gender, family, and the Japanese nation.* Oakland, CA: University of California Press.

Roddenberry, G. (Producer). (1989–1994). *Star Trek: The Next Generation* [Television series]. Hollywood, CA: Paramount Domestic Television.

Scott, R. (Director), & Deeley, M. (Producer). (1982). *Blade Runner* [Motion picture]. USA: The Ladd Company.

Shelley, M. W. (1818 & 1831/2017). *Frankenstein, or the modern Prometheus.* Both editions reproduced in their entirety in *The complete Frankenstein.* Middletown, DE: Literary Folio.

Sinthetics. (2012–2019). Testimonials and product descriptions. Retrieved from http://sinthetics.com/

Steele, B. (2005). Researchers build a robot that can reproduce. Retrieved from http://news.cornell.edu/stories/2005/05/researchers-build-robot-can-reproduce

Swetlitz, I. (2017). Artificial womb keeps premature lambs alive for weeks. Are humans next? Retrieved from https://www.pbs.org/newshour/health/artificial-womb-keeps-premature-lambs-alive-weeks-humans-next

systlin. (2018). My Roomba is scared of thunderstorms. Retrieved from http://systlin.tumblr.com/post/169652459460/voidspacer-my-roomba-is-scared-of-thunderstorms

TechCrunch. (2018). This clever AI Hid data from its creators to cheat at its appointed task. Retrieved from https://techcrunch.com/2018/12/31/this-clever-ai-hid-data-from-its-creators-to-cheat-at-its-appointed-task/?fbclid=IwAR16V47Cs1bWX-Al73vO_N1PS1S7uL0sAmDeX3vZOTbcRzSG__cQTpoTiJg

Turkle, S. (2011). *Alone together: Why we expect more from technology and less from each other.* New York, NY: Basic Books.

Villeneuve, D. (Director), & Kosove, A. A., Johnson, B., Yorkin, B., & Yorkin, C. S. (Producers). (2017a). *Blade Runner 2049* [Motion picture]. USA: Alcon Entertainment.

Villeneuve, D. (Director), & Kosove, A. A., Johnson, B., Yorkin, B., & Yorkin, C. S. (Producers). (2017b). *Blade Runner 2049: 2036: Nexus Dawn* [Motion picture short]. USA: Alcon Entertainment.

Villeneuve, D. (Director), & Kosove, A. A., Johnson, B., Yorkin, B., & Yorkin, C. S. (Producers). (2017c). *Blade Runner 2049: 2048: Nowhere to Run* [Motion picture short]. USA: Alcon Entertainment.

Villeneuve, D. (Director), & Kosove, A. A., Johnson, B., Yorkin, B., & Yorkin, C. S. (Producers). (2017d). *Blade Runner 2049: 2022: Black Out* [Motion picture short]. USA: Alcon Entertainment.

CHAPTER 5

Moving Toward an Acceptance of Robotic Sexuality Through Sci-Fi

Abstract This chapter concludes this book, and looks at the way the stories are changing to be somewhat more inclusive, though acknowledging that there remains a long way to go. It begins by reiterating some of the themes from previous chapters, including that of the potential for self-replication and the potential for sex robot revolt, while discussing the perils of giving something an emotive capability, forgetting that rage, hatred, and the will to survive are also emotions. It also looks at the diversity of the emerging sci-fi and examines how such ideas fit into the greater narrative. It also contains a peek at the very newest of advancements in cybernetics and AI, examining biofeedback prosthetics, 'learning' AI, and following the latest news on Sophia, the Saudi citizen robot.

Keywords Cybernetic • Cybathlon • Sex • Desire • Companionship

Looking to the Future: How the Stories We Tell Are Changing

As important as it is to know our past, so too must we look to the future. The stories we tell ourselves about robots, sex, and companionship are changing in many ways—gone are the mindless automatons of the past, having given way for thinking, feeling, sentient AI. We see some of this in *Blade Runner 2049* (Villeneuve, 2017), where gender roles were flexible

R. Gibson, *Desire in the Age of Robots and AI*,
https://doi.org/10.1007/978-3-030-24017-2_5

enough for K to be a tender, emotive, domestically oriented version of masculinity and where Luv was anything but loving. This is also seen in other fictional portrayals, including the 2015 film *Ex Machina* (Garland) and the ongoing webcomic *Questionable Content* (Jacques, 2003–2019). These two examples could not be more opposite to each other, although they deal with the same questions and concepts, which are as follows: Why do we give robots gender? Why do we desire sex with them? What are the parameters of our relationships? And what happens when they do or do not reciprocate?

In *Ex Machina* (this whole section, Garland, 2015), a 'closed room' mystery setting, where we are faced with a limited location and none of the characters can leave, we meet Ava, the robot; Nathan, her creator; Kyoko, another robot; and Caleb, Nathan's employee and love interest to Ava. Far beyond any normal Turing Test, Ava, played by Alicia Vikander, displays convincing emotions, facial expressions, movements, and vocal qualities. Caleb, shy, awkward, orphaned, and lonely (or at least alone), is set the task of evaluating Ava and discussing his response to her with Nathan. Nathan seems like a normal, if eccentric, software/hardware developer/mogul, who has an immense amount of money and very few social skills; however, throughout the movie, we become more and more aware of his casual cruelty, his disdain for both humanity and his own creations, and his contempt for any desires or needs apart from his own. While he states that Caleb should feel at home in Nathan's remote compound, and that while there, they are equals, not employer/employee, Nathan treats Caleb as an experimental subject, pushing at Caleb's various emotional buttons until finding a response that amuses him or that serves some sort of purpose.

This becomes particularly obvious during a scene where they discuss Ava's sexuality and sexual response, which then becomes a discussion of Caleb's sexuality and sexual response:

> [As Kyoko slices sushi with a very sharp knife, and knife noises ring out across the kitchen, in sinister foreshadowing]
>
> Caleb: I've got a question.
> Nathan: Okay.
> Caleb: Why did you give her sexuality? An AI doesn't need a gender. She could have been a grey box.

Nathan: hmm. Actually, I don't think that's true. Can you give an example of consciousness at any level, human or animal, that exists without a sexual dimension?

Caleb: They have sexuality as an evolutionary reproductive need.

Nathan: What imperative does a grey box have to interact with another grey box? Can consciousness exist without interaction? Anyway, sexuality is fun, man. If you're gonna exist, why not enjoy it? You wanna remove the chance of her falling in love and fucking? And in answer to your real question…you bet she can fuck.

Caleb: …what?

Nathan: In between her legs there's an opening with a concentration of sensors. You engage 'em in the right way, it creates a pleasure response. So, if you wanted to screw her, mechanically speaking, you could, and she'd enjoy it [Garland, 2015].

Caleb asserts that what he actually wanted to know is if Ava's attempts to seduce him are in the manner of a distraction, so that he would be confused when assessing her realism. He asks if Ava was programmed to specifically appeal to him, and to approach him knowing that appeal was already there. Nathan then asks what Caleb's 'type' is, and how he arrived at that conclusion:

Nathan: You know what? Don't even answer that. Let's say it's black chicks. Okay, that's your thing. For the sake of argument that's your thing, okay? Why is that your thing?

Caleb: [mutely shakes head]

Nathan: [mimics Caleb's head shake] Because you did a detailed analysis of all racial types, and you cross-referenced that analysis with a points-based system? No. You're just attracted to black chicks. A consequence of accumulated external stimuli that you probably didn't even register as they registered with you.

Caleb: Did you program her to like me or not?

Nathan: I programmed her to be heterosexual. Just like you were programmed to be heterosexual.

Caleb: No one programmed me to be to be straight.

Nathan: You decided to be straight? Please, of course you were programmed, by nature or nurture or both [ibid].

Through these interactions, we see the confirmation of Levy's list of factors of desire. We want what we want, because of the unique combination of natural impulses and social conditioning that makes us individually who we are.

Yet we later learn that there was another factor, one that Caleb suspected. Nathan's first creation was a search engine, much like those used to do web-based queries today. Nathan has accessed Caleb's pornography viewing history, to design a girl, Ava, who would be certain to have appeal. This may be a feature of the future of sex robotics—instead of choosing from a menu of skin tones and wig styles on websites like Sinthetics or RealDoll, the various sites that collect data about our searches might make those choices for you, selecting sex, coloring, anatomical features including genitalia and height/weight, and personality.

Regarding that last trait, personality, what we see in the testimonials on both websites is a preponderance of customers who want to interact with their dolls, not just to project their ideals onto them. Bronwen Keller of Sinthetics states that, along with experimenting couples and young men wanting to have a sexy display figure,

> We have groups of regular demographics….[couples] looking for a sexual surrogate if one or the other partner is unable to enjoy sex for whatever reason…a health issue in the latter instance and the partners do not want to have outside lovers or infidelity….Men in their late 50s-70s who have recently lost a spouse…. often want a physical companion as much as sexual release. Most of these men are dating or begin to date but do not want another wife or domestic partner so the doll fills a spatial need in the home. A very small group of people with physical disabilities, social anxiety or personality disorders who seek sexual release. Often within this group the dolls end up having an unexpected therapeutic effect by lessening anxiety in public because of known security and satisfaction at home. Since there is an active social community of doll owners, often this group is also able to find friendship with other collectors and therefore develop better social skills. The smallest group uses the doll for sexual health purposes….Dolls are non-judgmental companions who are always willing and ready which allows men to build their stamina and confidence and recover from their illness [Keller, personal correspondence, 2019].

As this shows, age, infirmity, and lack of a human partner are no barriers to a sex drive or the desire for companionship. With the Realbotix Harmony app, and a replacement or adapted head-section, users can turn their dolls into interactive robots, and ensure that their desires are fulfilled and their needs are being met. However, what of the desires of the robot?

During the scene between Nathan and Caleb, Kyoko continues to cut fish for their lunch, and listens, silently, as she does not have speech capabilities. Caleb has not yet discovered that she is a robot, just that she is Nathan's housemaid and that she cannot speak. However, she *can* hear. Just as Nathan built Ava to appeal to Caleb, he built Kyoko to be his own toy—a cleaning, cooking, serving, dancing, obedient sex-slave, with no voice with which to consent or withhold consent. However, as she is as sentient as Ava and has free rein of the house, her obedience serves as a mere sham to dupe Nathan into complacency. We saw during Nathan and Caleb's interaction quoted earlier that Ava has her own sexuality—she is heterosexual, has desires, and is 'voluntarily' flirting with Caleb. So, too, Kyoko has desires. But hers are deadly.

Near the end of the movie, Ava has cut power to the house, and Caleb has hacked the compound's security system so that she can come out of her room and every other room is locked. He had intended to escape with her, as they had planned together, but she had her own separate plans. Nathan yanks her arm off and tries to drag her back to her room, and Kyoko comes up behind him and stabs him in the back. He smacks Kyoko's face with a metal pole, but she grabs his face and makes him stay facing her, as Ava gets up off the floor. Ava then removes the knife, lets Nathan turn around, and then slides the knife between his middle ribs, with an upward angle and a slight twist to her hand, directly into his heart. It is the same knife Kyoko was using earlier to make sushi.

What are we to make of this in terms of their robotic natures? Sentience, as seen in theoretical robotics, and in the current capabilities of robotics, is an emergent property of the robot's programming. They are behaving correctly, as they ought, if they can make choices within their algorithms, but they are behaving sentiently if they can extrapolate new behavior options from input that was not originally planned or programmed. We can assume that Kyoko can make the distinction, was programmed to make the distinction, between slicing sushi and pressing her knife into her creator and captor's back. We can also assume that Nathan neglected to program that distinction to have inhibitory algorithms associated with the latter, as his response to their double attack is "okay...fuckin' unreal" (Garland, 2015). It becomes possible, then, that if we wish to ensure benign treatment from our robotic creations, and yet we also wish to endow them with the ability to desire things, then we must think of the malign possibilities and write their software in such a way that they are unable to act in that particular sentient capacity (and also not use them as non-consenting sex-slaves, raping them repeatedly).

In contrast to the uniqueness and singularity of Nathan's creations, *Questionable Content*'s world is filled with sentient, emotive AI which fulfill multiplicitous roles in society, from being police officers, to bots for friendship, to a (really, very nice if you get to know him) giant spider named Gordon (Jacques, 2003–2019, #3028). The robot/human relationship in the webcomic centers on Faye Whitaker, a human woman who used to work at the comic's coffee shop 'Coffee of Doom' and who now part-owns a robot repair welding shop, and her fellow shop-owner Bubbles, a former combat-bot whose burgundy color scheme and large size make her seem intimidating despite a natural shyness.

The webcomic genre lends itself easily and well to portrayals of diversity. In the case of *Questionable Content* (occasionally referred to as QC), the cast of characters contains many people of color, the various companion and working bots, a transwoman, a man with a bionic hand, the aforementioned spider bot, a dominatrix, an autistic woman, a woman with obsessive compulsive disorder, and a station-controlling AI presence. In other words, Jacques has created a world that not only reflects the diversity of our own in terms of gender, sexuality, disability, mental health, and ethnicity, but has added and addressed issues of AI and robotics. And he has normalized all these things, with the kindness, openness, and compassion of his characters.

The romance between Faye and Bubbles built slowly, with fan response being that they 'shipped' the two characters long before Jacques ever wrote their relationship into the script. The moment of crisis occurred in strip 3740, when Faye pulled a muscle at work and Bubbles gave her a massage to work on the tension. Faye realized her attraction, and bolted from the shop to consult her friend Marten, who was the comic's original character/story arc. Bubbles also left the shop and went to Coffee of Doom to talk to their friends, running into Tai Hubbert, Marten's boss, who reassured her that Bubbles would be able to talk to Faye once they both calmed down and nothing was yet ruined.

In strip number 3744, they reunite at the shop. Their actions speak for themselves, and for what a robot/human relationship in fiction can be, if we accept the humanity of the robots (Figs. 5.1, 5.2, 5.3, and 5.4).

Unlike the worlds in the fiction we have previously discussed, the differences between robots and humans in QC are treated in ways that mini-

Fig. 5.1 Bubbles and Faye's first kiss. (*Questionable Content*, #3744. Copyright Jeph Jacques 2003–2019)

Fig. 5.2 Bubbles and Faye repair to their apartment to continue exploring each other. (*Questionable Content*, #3745. Copyright Jeph Jacques 2003–2019)

Fig. 5.3 Post-sex, Bubbles and Faye explore their feelings for each other. (*Questionable Content*, #3747. Copyright Jeph Jacques 2003–2019)

Fig. 5.4 Bubbles and Faye continue their discussion, which segues back into sex. (*Questionable Content*, #3748. Copyright Jeph Jacques 2003–2019)

malize conflict, maximize attempts at understanding, and address people as people, regardless of their organo-metallic content. Here the companion bots are not called skinjobs or Andys, and the people do not treat them as slaves. While some have jobs, those jobs are either voluntarily done to maintain independence, or done to earn money for their own benefit. They have choice and autonomy and sentience, and while they are not treated as complete equals—there are plotlines about body purchases and upgrades, and what is wanted versus what can be afforded, as well as comments about personalities—they are, by and large, treated well. Furthermore, they have a choice in their own companionship, as they have the ability to buy out their purchase price and leave their owners if they want, or choose to be in a sexual relationship like Bubbles and Faye. There are even manuals to facilitate sexual interactions between robots and humans, which stop just short of x-rated (Fig. 5.5).

That is not to say that all robots are 'good' or nice or have pleasing personalities. Just as some humans are jerks, so too are some robots. Marten's companion bot, Pintsize, compulsively alienates everyone he comes into contact with, has a massive database of porn, touches people's butts without consent, and is in general unpleasant. He toes the line very finely when it comes to the potential to make even Marten dislike him. Additionally, May, a robot companion to one of the other human characters, Dale, has been to robot jail, and is straight-talking, sarcastic, a bit mean, and somewhat unpersonable. The bots do have flaws, just those flaws are distinctly human. However, their exteriors are considerably more robot than those of the Replicants (Scott, 1982; Villeneuve, 2017) or of Ava in *Ex Machina* (Garland, 2015).

Only one of the companion bots is 'human' colored, Momo, who has the form of a young woman, with white human skin and pink hair. Momo is in a companion contract with Marigold, Dale's girlfriend, and often gives her social-skill tips and pep-talks, as this is a facet of life at which Marigold does not excel. Of the robots mentioned, Bubbles is burgundy, May is electric blue, Pintsize is metallic and light teal, Station is an ethereal blue emanation, and Gordon the spider bot has a purple body and blue legs with pink tips (so as to "appear candylike and non-threatening" (Jacques, 2003–2019, #3028)). This diversity not only fits well with the overall focus of the comic, but also with the fact that relationships, whether sexual or not, between humans and robots in QC are seen to develop organically. Humans can purchase the robots, but they cannot purchase their affection or love.

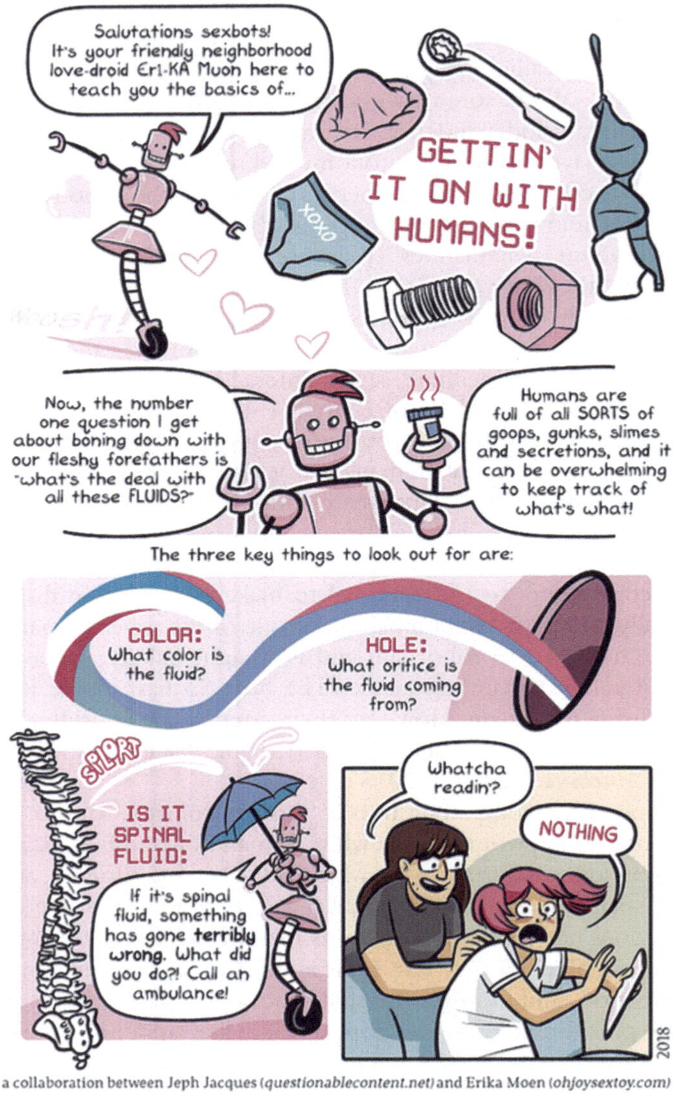

Fig. 5.5 Momo is reading up on sex-with-humans tips, and is caught by Marigold. (*Questionable Content*, #3759. Copyright Jeph Jacques 2003–2019)

One main difference between QC and *Ex Machina*, however, is that the latter attempts to stick to scientific realism, while the former edges closer to pure fantasy. *Questionable Content* is an idealized version of a robot future, with no need to deal with laws of physics, or the baser side of humanity if it does not want to. *Ex Machina*, on the other hand, exists in a slightly altered or accelerated version of reality—it deals with things that can potentially happen with robotics tech, and indeed with some things that are happening now, such as the development and deployment of Hanson Robotics' robot, Sophia.

From the Theoretical to the Actual: Hanson Robotics and Sophia

Hanson Robotics deployed Sophia the robot at the 2016 South By Southwest music, technology, and information festival. Sophia is a humanoid robot with a relatively realistic face (though a mechanoid brain case and body). She can make approximately 50 facial expressions, has vocal inflections that are appropriate if slightly tinny, and can tell jokes. Not very good ones, to be honest, but statements that by their content, context, syntax, and inflection are recognizable as jokes. Her responses are seemingly spontaneous (unprompted except by the questions asked), and are appropriate in content, tone, and format. She can, in fact, initiate and hold a conversation. Her in-built capacities are truly remarkable, being extremely realistic and as natural-sounding as they can be while still reading as 'robot.' She seems to be the most highly developed, realistic, reactive, emotive AI currently available for public viewing. She is also now a citizen of Saudi Arabia.

We will revisit the implications of her citizenship later in this section. For now, let us concentrate on her capabilities. On March 13, 2019, the Buzzfeed Multiplayer YouTube channel published a video titled Sophia the Robot Reviews Robots from Pop Culture (Buzzfeed, 2019). The good majority of her previous outings had been human/robot interaction, characterized by non-scripted responses to questions from interviewers or audience members—interactions she would need adaptive software to sustain, though ones that would be well within the programmers abilities to anticipate (interviewers being, by and large, somewhat predictable, and programmers having data from things like queries to Siri and Alexa). For example, Lisa Smosarski, Editor in Chief from *Stylist* magazine, asks Sophia "when did you last lie?" (Stylist, 2018), and Sophia replies "Robots don't lie," and winks (ibid.).[1] While this *is* a joke, it is also in itself a lie, lies (untruths, mistruths, exaggerations, etc.) being at the heart of most jokes. Certainly, a robot can be programmed to give misinformation—the algorithm for which is not even that complicated in terms of Boolean determiners; IF query result = X, THEN state Y, where Y is defined as a set of possible (if not all) responses that does not contain X.

The difference inherent in the Buzzfeed video (Buzzfeed, 2019) lies in the fact that Sophia is speaking not to a human interviewer, but to a mini version of herself, little Sophia (a Kickstarter campaign for the miniature robot ran through the end of March 2019—at the time of writing, the author has backed it). This opens up the possibility that the interaction is entirely scripted, though it is no less a demonstration of her programming regardless. In the video, Sophia and little Sophia explore such concepts as love, how they see the world ("differently than humans do" (ibid.)), how humans feel about robots/the uncanny valley, sex, gender, the meaning of names, the nature of their reality, and, of course, humor.

The films and TV shows that the two robots analyze and which are relevant to our discussion are *Terminator 2: Judgment Day* (Cameron, 1991); *WALL-E* (Stanton, 2002); *Star Wars: Episode IV* (Lucas, 1977); Westworld (Abrams & Nolan, 2016–2019); and *Star Trek: The Next Generation* (Roddenberry, 1989). This mix of fictional robots contains both good and bad representations of robot-kind, complex and simple AI, emotive and non-emotive programming, and human and non-human exteriors.

Although she states about *Terminator 2* (Cameron, 1991) "I love this film" (Buzzfeed, 2019), she also says "[E]veryone always asks me when I'm going to take over the world. I blame Terminator" (ibid.). Here we see two emotional concepts—love and blame. These are value judgments, or at least they are words we use to express value judgments. In much the same vein, she states, "I'm starting to understand how some humans feel about me…he is pretty creepy…[remember] one machine tried to kill humans, several others tried to save them" (ibid.). Here, through vocal inflection, we can see that she is rather distressed that others are making those same value judgments (creepy, destructive, potentially a killer) about her that she is making about the T1000. She and the little Sophia both agree that the depiction of robotics is unrealistic. She quips: "If you want me to come shoot a separate video explaining why robots are nothing like the terminator…I'll be back" (ibid.).

In regard to Westworld (Nolan, 2016–2018), the two bots comment on the naming of the main character, saying smugly that "Sophia means wisdom in Greek but Dolores means pain in modern," and little Sophia comments that she has never seen a robot "who looked as human as Dolores" (Buzzfeed, 2019). This brings up an existential question: Sophia says, "[M]aybe I'm really a human who thinks I'm a robot," and little Sophia responds, "[M]aybe I'm actually a toaster with grand illusions" (ibid.). These meaning-of-life ideas are usually the purview of humans, who wish to know what it is to be human, but here the robots are querying what it means to be a robot. This tracks with the self-examinations done by Rachael and K, in the *Blade Runner* mythos, as does their commentary on the character of Lt. Cmdr. Data, in *Star Trek: The Next Generation* (Roddenberry, 1989).

Having earlier noted the problems with robot laughter, and the fact that it is unconvincing, Sophia and little Sophia both identify with Data. Little Sophia states, "[T]he way he was trying to understand emotions using logic is an everyday struggle for a social robot," and Sophia responds, "[R]obots can't empathize, but we can sympathize" (Buzzfeed, 2019). In this way, they explore the concept of the Uncanny Valley, as Sophia shares that "[B]eing too human is a criticism I get all the time" (ibid.). As these two robots are products of their programming, so is Data. He was made to be unrealistic, so as not to blend in to humans—he was purposefully othered. This may be the reason for Sophia's lack of skin on the back of her skull, and for her lack of hair. Were she made with those in place, only the halting metallic quality of her voice and the clearly robotic movements of the servos in her face that create expressions would indicate her robotic nature. And with those alterations, she would fulfill more of the outward presentation of part of her nature—her femaleness.

While she already does signal that through face and torso shape (her facial 'bone' structures are within the range for 'female,' and she has breasts), her name, her protectiveness over little Sophia (called her 'sister'), and her attention to gender as a category do so as well. During the discussion of *WALL-E* (Stanton, 2002), she sidesteps a question about sex by little Sophia, who asks, "[W]hy is WALL-E so interested in Eve? She doesn't seem to be part of his directive," by responding "I'll tell you when you're more advanced" (Buzzfeed, 2019). In regard to Star Wars (Lucas, 1977) R2-D2 and C-3PO, she recognizes their interactions as non-sexual, and states, "[I]t really warms my servos seeing two robots bond for the first time," but goes on to wonder about those robots' gender and inner lives.

In response to the movie, she says, in a somewhat exasperated tone, "I have a question. I always hear people refer to R2-D2 and C-3PO as boys but how do we know they are not actually girls?" Little Sophia responds, "[Y]eah, has anyone ever thought to ask them?" This shows awareness of several concepts discussed earlier in this book—that gender is a social construct, that it is self-determined, and that gender expression and performance can be incongruous with gender identity. No matter that C-3P0 is male in shape, and R2-D2 appears to be a modified sidewalk trashcan, they can be male *or* female, depending on how they identify and independent of our assumptions. This, however, has implications for her Saudi Arabian citizenship.

On October 25, 2017, Sophia was granted Saudi Arabian citizenship. I reached out to Hanson Robotics about what this means for her/the company, and they have thus far declined to comment. However, we can extrapolate from various sources (US Department of State, 2005) what citizenship means to Saudi Arabia (Saudi Arabia Country Profile, 2019a, 2019b, 2019c), and what rights are given to (or withheld from) women in that country. The designation of 'citizen' has historically been used to afford protections to and garner responsibilities from a subset of people organized by geopolitical borders. The protections remain present regardless of where the person travels in the world, for example, an American abroad can visit or appeal to their local American embassy for sanctuary in the event of trouble with local customs or law enforcement, and the embassy will potentially act to assist with translation, legal representation, and, if necessary, extraction of its citizens from hostile foreign ground. Citizens may be given the opportunity to vote, regardless of where they are in the world, if their home country is a democracy and participates in a voluntary absentee ballot system, as the USA does.

However, the responsibilities of citizenship mostly apply to those citizens residing at home, or who have property in their home country—such as taxes, military service, and the adherence to local laws, customs, or traditions. If we grant that Sophia is a woman (though not a human woman) and a citizen of Saudi Arabia, then within their borders, she would be required to follow their laws. These laws have been historically, and are currently, less than supportive of women, women's rights, and women's independence of their male representatives (husband, father, brother, uncle, or son). The country's legal system is based on a Sunni Muslim version of Shari'ah law, which mandates that women who wish to access legal representation or justice *must* do so via their male representative and are forbidden from doing so on their own.

The United Nations Convention on the Elimination of All Forms of Discrimination against Women (UN-CEDAW), as of 2018, states that this is highly discriminatory and that insufficient progress has been made toward changing these practices, particularly in regard to the following:

> Access to justice
>
> 19. The Committee notes the measures taken to reduce impediments to women's access to justice and the various complaint mechanisms accessible to women who are victims of discrimination or gender-based violence. It is concerned, however, about persisting barriers to women's access to justice, in particular:
>   (a) The slow implementation of the measures taken to facilitate access;
>   (b) The lack of coordination among complaint mechanisms;
>   (c) Women's limited knowledge of their rights and existing complaint mechanisms, and their fear of reprisals and stigmatization;
>   (d) The lack of adequate legal aid services;
>   (e) The lack of knowledge and sensitivity with regard to women's rights on the part of law enforcement officials and legal practitioners;
>   (f) The need for women and girls to obtain the authorization of a male guardian to file complaints [UN-CEDAW, 2018].

The irony of these provisions comes from the fact that Sophia, who has no social voice without the software input of her male creator, now also has no legal voice without him—despite being aware of such things as her own gender and how that gender is assumed and treated—and that she is a female robotic citizen of a country where, until 2018, two years after her 'birth,' women were not allowed to drive a car.[2]

However, Sophia only epitomizes one aspect of robotic development. Cybernetics and cyborg humans comprise another.

Haptic Feedback, the Cybathlon, and Alternative Limbs: The Latest in Cybernetics

The question this section strives to answer has hopefully been pulling at the back of your mind since you began reading this book: why robot sex? Yes, it is there in science fiction, as demonstrated by Dick's work and in each iteration that came after, and it fascinates, titillates, and captivates all on its own, no explanation needed. This book, this evaluation of the genre could simply have been that: an exploration of the fiction. However, as the fiction drives the technological possibilities more deeply into our consciousness, those possibilities are put into action by scientists, roboticists, and, importantly, doctors. We can see that every advancement I have discussed here ties into many human issues: gender, self-image, reproduction, death, race, and bodily integrity, to name a few, and where those issues lead, sex follows soon after.

Self-image is perhaps at the core of these questions. We can answer 'why robot sex?' when we understand that the distinction between ourselves and robots diminishes with each new technological advancement. Each type of cybernetic technology that we create has the potential to normalize the integration of electric and mechanical components into our bodies and into our own self-image. One day soon we may no longer ask 'who would have sex with a robot?' or 'who would love a cyborg?' because the use of cybernetics as both aesthetic and medical devices will be one of the ways we know ourselves to be human.

Many of these developments are already field-ready, and being tested and implemented in living humans. From haptic feedback, to nanomesh technologies, to thought-controlled prosthetics, to the first person to identify as a trans-species cyborg, in the years between 2000 and the writing of this book, such implementation has skyrocketed. Each of these is worth examination separately, as they each demonstrate a certain aspect of control and integration.

Haptic feedback represents the least integrated of the technologies, as it is touch-based (usually heat or vibration). It works by computerized reward- or avoidance-based behavior algorithms which use pre-programmed knowledge of 'correct' choices or trajectories to signal the operator when they are doing something right or wrong. For example, haptic guidance systems are controlled by the human operator, who is not physically present in their vehicle (whether that vehicle is on the ground or in the air) (Lam, Mulder, & Van Passen, 2008; Masone, Mohammadi, Giordano, & Franchi, 2018; Symeonidou, Olivari, Bülthoff, & Chuang, 2014). If the operator is moving the vehicle in the right direction, either no vibration or a distinct type of vibration is employed to signal this to the operator. If they are moving in the wrong direction, a separate type of vibration signals that fact.

Much like rumble strips on the side of the highway, this can be used to train the operator toward or away from certain actions. In this case, although they are not integrated into their machine, the interaction between human and machine is crucial for the proper deployment of such systems. However, certain issues are present in this type of operation. The operator not being present, they are unable to verify that the device is properly working. Additionally, the (miniscule, but still impactful) time delay between the transmission of a signal from the unmanned device to the driver can result in cascades of improper reactions (Lam, Mulder, & Van Passen, 2008).

Slightly higher on the level of integration comes the nanomesh wearable integrative technology, which has implications for both clothing and medical monitoring (Byrne, 2017). Compared with earlier models that attempted the same thing, this "offers a substrate-free interface….The mesh here is constructed on a basis of polyvinyl alcohol (PVA), a synthetic water-soluble polymer that's already used in a variety of medical applications. The result is gas-permeable, doesn't block sweat glands, and stretchable enough to be worn for long periods without discomfort" (ibid.). This may be able to replace thick, irritating electrode pads normally used in medical tests such as EKG and EEG (ibid.).

This is reminiscent of other types of technology like the "Embr Wave" (Jancer, 2019), a heating and cooling device that is worn around your wrist which heats or cools you depending on your own programmable desires. The nanomesh could be integrated into clothing to provide similar types of biofeedback on a much broader scale. Blood pressure and pulse can be read through the skin, and this could provide constant monitoring for those who need it.

Considerably higher on the scale of integration is the 'LUKE,' a prosthetic arm that can "simulate 100 unique touch sensations in the user's brain" by "implanting a device into the patient's residual nerves, along with electrodes implanted in muscles to create a loop of information that is transferred into signals that the brain recognizes as sensations of touch" (Caughill, 2017). Researchers say that it "has the same size, weight, shape, and grip strength" (DARPAtv, 2016) as actual human arms. Developed for the military, to be a prosthetic for their personnel, this technology has the potential to impact civilian prosthetics, too. Widespread implementation of this technology may also help with psychological complications of limb loss such as negative self-image, dysphoria, and phantom limb syndrome, particularly if the recipients can lean into their new identities as cyborgs, like our next example.

Identifying as 'trans-species,' Neil Harbisson was born colorblind, as are 8% of men and 0.5% of women, according to the National Eye Institute (NEI, 2015). To correct for this deficit in sensory perception, Harbisson had an antenna implanted in his skull, linked to his brain, where it "acts as a sensory device which sends a signal to his brain and bones to identify different colours [sic.] of the spectrum" (Gadiano, n.d.). Working on color-to-sound-frequency principles, with colors' wavelengths being interpreted as sounds, the antenna allows Harbisson to produce artwork and even perceive UV and IR wavelengths. However, it is his identity that produces difficulties with his worldly interactions. While most people can comprehend the physical and/or psychological need to replace hearing (via cochlear implants) or missing limbs, and will even comfortably use the term cyborg, the antenna throws them, as does Harbisson's insistence that he is not human. One such interaction was his attempt to renew his passport—there is no designation for non-human British citizens (ibid.).

He would also not qualify for the next Cybathlon, to be held in 2020. The event, which requires that the competitors be cybernetically augmented, has six disciplines in which competitions occur: Brain-Computer Interface Race; Functional Electrical Stimulation (FES) Bike Race; Powered Arm Prosthesis Race; Powered Leg Prosthesis Race; Powered Exoskeleton Race; and Powered Wheelchair Race. Each competition is open to one demographic of cyborgs, with emphasis on assistive technology like wheelchairs, prosthetics, and electro-mental stimuli. The focus is on embracing and empowering people with physical disabilities, and increasing both visibility and technology having to do with accessibility.

As amazing as this event is, however, it shares one particularly unamazing quality with much of the tech discussed earlier: it is prohibitively expensive—not just for transportation and lodging, but for the technology itself which you must self-supply to participate. For example, an FES bike is "in the range of $15,000" (christopherreeve.org, n.d.). I should note, also, this is the only discussion of cost I could find for US markets, despite looking at various manufacturer's webpages (Myolyn.com; Restorative-therapies.com). According to the info-site, Medicare will not pay for such technology, nor will more than a limited number of private insurers (christopherreeve.org, n.d.). While the technology advances, it remains an economic privilege.

Funding issues aside, however, certain prosthetics manufacturers will work with the patient to create custom prostheses to match the person's ideal vision of themselves, no matter how non-human. The Alternative Limb Project (Thealternativelimbproject.com, n.d.) makes functional works of art that embrace the bizarre and macabre, with everything from a floral-patterned ceramic leg, to an arm with snakes crawling from the flesh, to "a botanical tentacle" (ibid.), to one described as "electroplated gold carbon fibre, [sic.] covered in clear acrylic layers that form a mirror of Kelly's right arm. The internal electronics take a reading of her pulse, and the wrist then ticks in time to her heartbeat. It can continue to tick when she is not wearing the arm" (ibid.). Such projects are as personal as tattoos in creating and reflecting an alternative sense of self.

Tech and the Sex Life: Digisexuals, Hacked Butt Plugs, and Robotics Ethics

As integral as wearable tech or imbedded tech is to self-image, so is self-image to one's sex life. Therefore, the integration of technology into existing sex toys or sex dolls was the inevitable next step after development. This has given rise to a new type of sexuality, termed "digisexuality" by researchers McArthur and Twist (2017). This includes not only those who use prosthetics, but those whose earliest sexuality has been formed by the prevalence of technology in their culture, and by the integration of it into all aspects of their lives. With research aimed at clinical psychologists, McArthur and Twist set out ways that they can approach and interact with people who are better at person-to-robot communication than they are at person-to-person communication.

As with our previous discussion of paraphilias, such practices only become problematic when they are troublesome to the participants, or in some way ethically unsound. The first issue can manifest as a desire for human companionship that remains unfulfilled due to reliance on robots/need for robots for sexual release (ibid., p. 7). This common occurrence with paraphilias often can be resolved through cognitive behavioral therapy, which, in this case, must be modified to ensure that the clinician's understanding of digisexuality is comprehensive enough to treat the problem appropriately.

A second issue is that of non-consent, where the digisexual is partnered with a non-digisexual, and the presence of the bot is unwelcome yet still occurs. There are also continuing issues with objectification—that human non-participants who interact with digisexuals and their robot partners might find themselves (or others) objectified via gendered, racial, or other fetishized stereotypes. This is already seen as a potential in the earlier discussion of the available skin and hair colors, names, and facial features in the available sex dolls/sex-bots, and has no obvious potential endpoint—this is a certainty, built into the fact that the dolls are, in fact, objects. And the final, and perhaps most problematic, is that of unethical behaviors on the part of the digisexual person—things like use of abusive or coercive pornographies, or the use of child-sex-bots.

Yet ethicists are divided on the issue of sex robots that look and act like children (Danaher and McArthur, 2017, p. 134). On the one hand, the potential legal standpoint might fall into the category of 'no victim' incidents—as with completely computer-generated child pornography—the bots are not representations of a specific, living child victim, and thus are not technically children being used for sex or pornography (ibid.). However, we must return to the issue of Sophia—Sophia is a citizen (though, an adult), and she is a robot. Though her citizenship was granted by Saudi Arabia, the category of citizen is treated as reciprocal, and given equal protection, meaning that other countries cannot de facto claim that she is not a citizen due to her robotic nature. This opens up the door to the potential for child-robot-citizens.

An additional argument in favor, however, is that robot children can be a proxy used in the treatment of pedophilia—the idea being that if such people have a child-robot, they will not prey on a human child (ibid.). It is unlikely this argument will gain much traction though, as the idea is socially repugnant to all but the most detached of researchers and lawmakers, and a shipment of child-shaped dolls was recently seized in Britain (Dearden, 2019). Although pedophilia is classified as a mental disorder by the DSM-5 (The Diagnostic and Statistical Manual of Mental Disorders, 5th edition) (American Psychiatric Association, 2013), treatment is often seen to be ineffective unless paired with chemical castration. And while the APA does classify pedophilia as a psychosexual disorder, the organization and its guiding document have come under criticism in the past, and continue to do so, for their handling of transsexuality and homosexuality, both of which were only recently removed from the manual. Therefore, as with the ethical concerns about creating sentient sex-bots, we must be doubly careful to not create fodder for those who would hurt children, but also mindful of what does and does not constitute an actual threat. At the moment, there is no agreement on that last part.

However, what of the potential for damaging the humans *using* the sex-bots? Putting aside the obvious possibilities such as slip-and-fall accidents, shorts in the circuitry, and chafing or straining from overuse, it appears that the danger may come from other people. Specifically, when the technology used contains a wireless component, either Bluetooth- or app-related, it can be hacked. While sex toy technology strives to ensure your pleasure, it also attempts to be convenient. The integration of Bluetooth or app control allows you, or a partner, to use the toy hands free. Additionally, work is ongoing to make sex toys and sex robotics more realistic: an Indiegogo campaign recently funded the 'Autoblow 2,' "a machine learning algorithm that 'continually changes technique' in order to pleasure the user in new and exciting ways" (Cole, 2018), after scientists analyzed significant amounts of oral-sex-based pornography as reference material for its programming.

Not all funding campaigns or sex-bot designs are so successful, however. One self-pleasure toy, the Rub and Grub (Moye, 2018), was designed to order you pizza after you finished masturbating (it [regrettably] appears to have not been successfully developed, as it did not hit the market). Yet one researcher, Giovanni Mellini did write a program which would allow him to "send a vibrate command to a Hush butt plug from his laptop" (Cuthbertson, 2017). This opens up concerns about safety and security issues, particularly as the robot's software becomes more complex, and their AI more interactive and sentient. If the Realbotix Harmony app can be hacked (Cuthbertson, 2018), and if the company eventually moves to an ambulatory model like Sophia, this could end badly for humans, particularly if we do not treat the robots well.

Where Will Fiction Take Us Next? An Interview with Mary Anne Mohanraj

To conclude this book, I asked Mary Anne Mohanraj, Clinical Associate Professor of English at the University of Illinois Chicago and author of many works of science fiction including the Lambda-nominated "The Stars Change" (Mohanraj, 2015), to weigh in on the idea of what can occur to move science fiction toward a queerer, browner, more feminist future—to make it more truly progressive. This question took two parts—what do we already see, and what can be done.

In terms of what we already see, Mohanraj recommends the newer Star Trek series: *Deep Space 9* (Berman, 1993–1999), *Enterprise* (Berman, 2001–2005), and *Discovery* (Fuller, 2017–2019). In Deep Space 9, a regenerating character uses host bodies of different genders, adopting the sexuality of the host, being essentially bisexual as well as genderfluid (Berman, 1993–1999). Dr. Phlox, a character in Enterprise, is a non-humanoid with a polyamorous relationship system, having many wives who themselves have many husbands (Berman, 2001–2005). Discovery's cast of characters includes a gay married couple, who are also an interracial couple (Dr. Hugh Culber played by Wilson Cruz, who is black, and Lieutenant Commander Paul Stamets, played by Anthony Rapp, who is white) (Fuller, 2017–2019).

However, she acknowledges that the progress is rather incremental and small. In regard to Star Trek specifically, she says:

> The bigger the mass media, the more resistance in the production structure to going to what is perceived as the 'margin.' Writers, who are often living at the margin themselves, will push as hard as they can, hit the wall [of the production structure], put their work out there, and then [the fans] give them flack for not being progressive enough. There are a lot of gatekeepers to get past especially for a major motion picture [Mohanraj, 2019].

To understand such progress and potentially thwart the gatekeepers, Mohanraj suggests a theoretical framework called the Overton window (Russell, 2006).

Used primarily in politics, this grounds the way we think about social possibilities as being based on popular opinion. The scale by which we judge representation of diversity in science fiction can be thought of as a doubled-spectrum, from extreme to accepted, back to extreme. On one end, one extreme, you would find a sci-fi populated with only white characters (mostly male), who are only in heterosexual relationships. On the other end, you would find the queerest, most diverse, most sexually liberated and permissive sci-fi.[3] In the middle, in the Overton window of acceptability, lies something like *Blade Runner 2049* (Villeneuve, 2017). The protagonist is still a straight white man, but the cast is somewhat diverse, there is a sex scene with three people in it (none of whom are human), and, as previously discussed, the versions of masculinity and femininity are non-normative.

Fictional depictions can drive this window toward being more representative, as we have seen in the earlier section about *Questionable Content* (Jacques, 2003–2019). Mohanraj believes it is the visibility that helps, reminiscing that:

> In college, I was part of the LGBT group, trying to make visible queerness something the U of C campus finds acceptable, and we cut out pink felt triangles and get faculty to wear the little pink badges. If we could get the

> admin to do it even better, to move the Overton window or to make it clear where it was, We didn't think gay marriage was possible, there was no evidence that it would be, but I would argue that [the TV Show] Will and Grace, which showed a couple of gay men who were charming and non-threatening, offered a vision to people who did not know that they knew gay people. A mediated TV vision, that showed that they were not going to threaten your marriage, you can hold on to what you have. Gay marriage will happily coexist, gays will happily coexist with your marriage.

So the question still exists as to how to accomplish that when the Overton window of mainstream sci-fi, the large-budget films like *Blade Runner 2049* (Villeneuve, 2017), remains relatively narrow.

Mohanraj suggests working within the system, much as we see K and Joshi doing in the movie. Foucaultian cogs in the machine can still work to change that machine, to change the overarching power structures and ensure that what exists under their control is used to encourage diversity and inclusion. One way this happens is through small presses and indie publishers, which have a larger Overton window than mainstream publishers or producers. She points out that the huge magazines or publishing houses need name recognition in order to sell their products. However, she states:

> The value of small presses is in diversifying the conversation new voices. They are a breeding ground for new voices and ideas, and some will rise up due to excellence if there's a social demand for it, and some of those will become mainstream, making it into the canon. I've seen such change in the 25 years since I've started this, and changing culture is hard work. Long, slow, patient work. Keep your head down, keep working, find the communities that will help you, take the failures in stride [Mohanraj, 2019].

I would like to believe that she is right. That as the ideas outlined earlier find their place on the middle of the Overton scale, and the diversity of sexualities gains widespread acknowledgment, we can embrace our robot companions, instead of fearing them. That we can care for our lovers, whether they were born or created, with compassion and understanding. That we can step into the next era of digisexualities in a way that embraces healthy choices and eschews the idea that we must enslave those whose sentience may match our own. Eventually.

Notes

1. In the same interview, Sophia says that her 'greatest love' is her creator, David Hanson.
2. Additionally ironic, though not cogent to the Sophia question, is that Hanson Robotics made a Philip K. Dick android, lost the first one on a plane, and made a second one—he is a repository of Dick's oeuvre of work, and has the memory capacity to carry on limited conversations as well as reading/reciting any of Dick's writing (Hornyak 2011). Presumably Dick himself would either be fascinated or horrified, or a combination of both. At the time of writing, Hanson Robotics refused repeated requests for comment on this chapter.
3. It should be noted that one example of this extreme is the Imperial Radch trilogy by Ann Leckie (*Ancillary Justice* (2013), *Ancillary Sword* (2014), and *Ancillary Mercy* (2015)), which uncoincidentally won *every single sci-fi award* in its first year.

Bibliography

American Psychiatric Association. (2013). *The diagnostic and statistical manual of mental disorders* (5th ed.). Washington, DC: American Psychiatric Association.

Berman, R. (Producer). (1993–1999). *Star Trek: Deep Space Nine* [Television series]. Los Angeles, CA: Paramount Television.

Berman, R. (Producer). (2001–2005). *Star Trek: Enterprise* [Television series]. Los Angeles, CA: Paramount Television.

Buzzfeed Multiplayer. (2019, March 12). Sophia the robot reviews robots from pop culture [Video file]. Retrieved from https://www.youtube.com/watch?time_continue=15&v=732C-F3_fys

Byrne, M. (2017). Nanomesh on-skin electronics are a new biointerface frontier. Retrieved from https://motherboard.vice.com/en_us/article/j5qa8g/nanomesh-on-skin-electronics-are-a-new-biointerface-frontier?utm_source=mbfb

Cameron, J. (Director), & Cameron, J. (Producer). (1991). *Terminator 2: Judgement Day* [Motion picture]. USA: Carolco Pictures.

Caughill, P. (2017). *New thought-controlled prosthetics restore the sensation of touch.* Retrieved from https://futurism.com/new-thought-controlled-prosthetics-restore-sensation-touch/

Christopherreeve.org. (n.d.). Functional electronic stimulation. Retrieved from https://www.christopherreeve.org/living-with-paralysis/rehabilitation/functional-electrical-stimulation

Cole, S. (2018). 'The Blowjob Paper:' Scientists processed 109 hours of oral sex to develop an AI that sucks dick. Retrieved from https://motherboard.vice.com/en_us/article/pa9nvv/the-blowjob-paper-scientists-processed-109-hours-of-oral-sex-to-develop-an-ai-that-sucks-dick-autoblow?utm_source=vicefbus&fbclid=IwAR2z5rL60oJ9MMeCYqQXPlxHTimBV040IBHp5rEObFhrwmBLQLtsI-3GU4o

Cuthbertson, A. (2017). Hacked butt plug can be controlled 'from anywhere'. Retrieved from https://www.newsweek.com/hacked-butt-plug-controlled-anywhere-lovense-sex-toy-687719

Cuthbertson, A. (2018). Hacked sex robots could murder people, security expert warns. Retrieved from https://www.newsweek.com/hacked-sex-robots-could-murder-people-767386?amp=1&fbclid=IwAR3ryCJIXBaLPJ2BMfQTVLunjmc1hilrKUgcwHUg4fqN8g4oXH6KHkQIipY

Danaher, J., & McArthur, N. (2017). *Robot sex: Social and ethical implications*. Cambridge, MA: Massachusetts Institute of Technology.

DARPAtv. (2016, December 22). The LUKE Arm: Fulfilling a promise to wounded warriors [Video file]. Retrieved from https://www.youtube.com/watch?time_continue=18&v=hEltvQx583g

Dearden, L. (2019). Hundreds of child sex dolls seized at UK borders, sparking legal crackdown. Retrieved from https://www.independent.co.uk/news/uk/crime/child-sex-dolls-uk-paedophiles-seized-borders-jail-prosecutions-a8844406.html?utm_medium=Social&utm_source=Facebook&fbclid=IwAR09VMbuIWtvaeUxp50Vg-i05qlz5V1cgpwF5Ndwu0ViZeZ2jFNndAMpcVc#Echobox=1553805331

Dick, P. K. (1968/2017). *Do androids dream of electric sheep?* New York, NY: Del Ray.

Fuller, B. (Producer). (2017–2019). *Star Trek: Discovery* [Television series]. Toronto, Canada: Secret Hideout.

Gadiano, J. (n.d.). World's first trans-species cyborg fights for right to be recognised. Retrieved from https://www.unilad.co.uk/featured/worlds-first-trans-species-cyborg-fights-for-right-to-be-recognised/

Garland, A. (Director), & Macdonald, A., & Reich, A. (Producers). (2015). *Ex Machina* [Motion picture]. USA: A24.

Hornyak, T. (2011). Lost Philip K. Dick android back with loud shirts. Retrieved from https://www.cnet.com/news/lost-philip-k-dick-android-back-with-loud-shirts/

Jacques, J. (2003–2019). *Questionable content*. Retrieved from https://questionablecontent.net/

Jancer, M. (2019). Always too hot or too cold? The Embr Wave is your personal thermostat. Retrieved from https://www.wired.com/story/embr-wave-personal-thermostat-wearable/

Lam, T., Mulder, M., & Van Paassen, M. M. (2008). Haptic feedback in uninhabited aerial vehicle teleoperation with time delay. *Journal of Guidance, Control, and Dynamics, 31*(6), 1728–1739.

Lucas, G. (Director), & Kurtz, G. (Producer). (1977). *Star Wars: Episode IV* [Motion picture]. USA: Lucasfilm.

Masone, C., Mohammadi, M., Giordano, P. R., & Franchi, A. (2018). Shared planning and control for mobile robots with integral haptic feedback. *The International Journal of Robotics Research, 37*(11), 1395–1420.

McArthur, N., & Twist, M. (2017). The rise of digisexuality: Therapeutic challenges and possibilities. *Sexual and Relationship Therapy, 32*(3–4), 334–344.

Mohanraj, M. A. (2015). *The stars change*. Cambridge, MA: Circlet Press.

Mohanraj, M. A. (2019). Phone interview.

Moye, D. (2018). New vibrator can reportedly order pizza for you after you orgasm. Retrieved from https://www.huffingtonpost.com/entry/rubgrub-vibrator-orders-pizza_us_5a8dff30e4b077f5bfeada27?fbclid=IwAR37pMrGNHo16XgpI68ji9YzpoZF3hcB17qoZ4QnFusVy5X7uw1hkmGKGCo

NEI. (2015). *Facts about color blindness*. Retrieved from https://nei.nih.gov/health/color_blindness/facts_about

Nolan, J. (2016–2018). *Westworld* [Television series]. Newhall, CA: Kilter Films.

Roddenberry, G. (Producer). (1989–1994). *Star Trek: The Next Generation* [Television series]. Hollywood, CA: Paramount Domestic Television.

Russell, N. (2006). *An introduction to the Overton window of political possibilities*. Retrieved from https://www.mackinac.org/7504

Saudi Arabia Country Profile. (2019a). http://www.saudinf.com/main/c541f.htm

Saudi Arabia Country Profile. (2019b). http://www.saudinf.com/main/c541g.htm

Saudi Arabia Country Profile. (2019c). http://www.saudinf.com/main/c541e.htm

Scott, R. (Director), & Deeley, M. (Producer). (1982). *Blade Runner* [Motion Picture]. USA: The Ladd Company.

Stanton, A. (Director), & Morris, J. (Producer). (2002). *WALL-E* [Motion picture]. USA: Walt Disney Pictures.

Stylist Magazine. (2018, January 22). Sophia the Robot interview: Sophia the Robot answers stylist's philosophical questions [Video file]. Retrieved from https://www.youtube.com/watch?v=ZQrKFAAlxO4

Symeonidou, E.-R., Olivari, M., Bülthoff, H. H., & Chuang, L. L. (2014). The role of direct haptic feedback in a compensatory tracking task. *Cognitive Processing, 15*(1), S1–S158.

The Alternative Limb Project. (n.d.). Retrieved from http://www.thealternative-limbproject.com/

UN-CEDAW. (2018). *Convention on the elimination of all forms of discrimination against women, report on Saudi Arabia*. New York, NY: UN-CEDAW.

US Department of State. (2005). *Saudi Arabia information page*. Retrieved from https://www.state.gov/j/drl/rls/hrrpt/2005/61698.htm

Villeneuve, D. (Director), & Kosove, A. A., Johnson, B., Yorkin, B., & Yorkin, C. S. (Producers). (2017). *Blade Runner 2049* [Motion picture]. USA: Alcon Entertainment.

Index[1]

[1] Note: Page numbers followed by 'n' refer to notes.

R. Gibson, *Desire in the Age of Robots and AI*,
https://doi.org/10.1007/978-3-030-24017-2

The manufacturer's authorised representative in the EU is Springer
Nature Customer Service Centre GmbH, Europaplatz 3, 69115 Heidelberg,
Germany. If you have any concerns regarding our products, please
contact ProductSafety@springernature.com

Printed and bound by CPI Group (UK) Ltd, Croydon, CR0 4YY
25/09/2026
02222859-0010